Henry Hudson Nicholson

Laboratory Exercises

With Outlines for the Study of Chemistry...

Henry Hudson Nicholson

Laboratory Exercises
With Outlines for the Study of Chemistry...

ISBN/EAN: 9783337275822

Printed in Europe, USA, Canada, Australia, Japan

Cover: Foto ©berggeist007 / pixelio.de

More available books at **www.hansebooks.com**

LABORATORY EXERCISES

WITH

OUTLINES FOR THE STUDY OF CHEMISTRY

TO ACCOMPANY ANY ELEMENTARY TEXT

BY

H. H. NICHOLSON
Professor of Chemistry in the University of Nebraska

AND

SAMUEL AVERY
Professor of Chemistry in the University of Idaho

NEW YORK
HENRY HOLT AND COMPANY
1899

Copyright, 1899,
BY
HENRY HOLT & CO.

PREFACE.

THE primary object of the authors in preparing this book is to place in the hands of the science teachers in the State a laboratory manual adapted to the average high-school requirements. In it we have endeavored to emphasize the value of laboratory instruction *in the laboratory.*

The aim is to give the student facts, practically of his own finding, before a discussion and correlation of these facts. Our experience in teaching has led us to the conclusion that to reach the best results a student of chemistry must first be given experimental work. In this way his interest is sufficiently aroused to hear with profit descriptive lectures, or to read with some attention the discussion of facts and principles as found in elementary text-books. To this end he is carefully guided in his experimental work, after which he is expected to study the subject of his experiment in one or more of the texts referred to at the end of each exercise.

These texts are not necessarily the best that might be given, but they are in many cases the only ones available to the pupil. They are in all cases those found to be at hand in class use or otherwise in the high schools of this State.

In all cases the references themselves are to the specific subject under investigation.

It is not intended that this manual should take the

place of any of the many excellent text-books now in use, but rather that it shall supplement and make more useful those now in the hands of the pupils or in the school libraries.

The authors' acknowledgments are due to Mr. Jesse E. Beans of Omaha for his efficient service in preparing the illustrations. We would also express our thanks to Dr. John White of this university for valuable suggestions.

THE AUTHORS.

UNIVERSITY OF NEBRASKA,
June 10th, 1899.

ABBREVIATIONS.

THE abbreviations calling for explanation are as follows:

cc. cubic centimeter. grm. gram.
cm. centimeter. mm. millimeter.

Under "References":

Clk. Elements of Chemistry, by F. W. Clarke. New York: American Book Company. Date of copyright 1884.

Cly. New Text-book of Chemistry, by LeRoy C. Cooley. New York: American Book Company. Date of copyright 1881.

R. (Briefer Course.) An Introduction to the Study of Chemistry, by Ira Remsen. New York: Henry Holt and Company. Date of copyright 1893.

S. Elements of Inorganic Chemistry, by James H. Shepard. Boston: D. C. Heath and Company. Date of copyright 1885.

S. and L. An Elementary Manual of Chemistry, by F. H. Storer and W. B Lindsay.

	New York: American Book Company. Date of copyright 1894.
W.	Elements of Chemistry, by Rufus P. Williams. Boston: Ginn and Company. Date of copyright 1897.

SUGGESTIONS.

A NUMBER of carefully selected reference books should be placed in the school library. The following are especially recommended:

    Inorganic Chemistry (Advanced Course), by Ira Remsen. Henry Holt and Company, New York.

    General Inorganic Chemistry, by Paul C. Freer. Allyn and Bacon, Boston.

    Organic Chemistry, by Ira Remsen. D. C. Heath and Company, Boston.

    Treatise on Chemistry, by Roscoe and Schorlemmer. Volumes I and II. D. Appleton and Company, New York.

The articles on chemical topics in the "Britannica" will be found especially valuable. The lives of famous investigators may be studied in connection with their discoveries, thus:

In connection with	The life of
Oxygen	Priestley.
Chlorine	Scheele.
Air	Cavendish.
Atomic Weights	Berzelius and Stas.
Atomic Theory	Dalton.
Combustion	Lavoisier.
The Alkali Metals	Davy.

TO THE PUPIL.

Work independently. You are not concerned with what your fellow student is doing.

Do not talk, except with your instructor about your work.

Do not ask questions of your instructor until you have tried to answer your questions for yourself.

Keep in a note-book, especially reserved for this use, an accurate record of your laboratory work. Record your observations when you make them, not from memory afterwards.

Do not hurry.

Keep the apparatus clean.

Work cautiously, using small quantities of reagents; avoid inhaling poisonous gases, and be especially careful to avoid getting acids, alkalies, or any other corrosive or poisonous substance into the eyes.

CONTENTS.

	PAGE
ABBREVIATIONS	v
SUGGESTIONS	vii
TO THE PUPIL	viii

THE NON-METALLIC ELEMENTS.

EXERCISE
1. Preliminary Chemical Manipulation 3
2. Physical Changes and Chemical Changes 7
3. Conditions under which Chemical Changes take place 9
4. Elements. Mixtures and Compounds 11
5. Some Chemical Terms 13
6. Oxygen: Preparation and Physical Properties 15
7. Chemical Properties of Oxygen. Problems 18
8. Hydrogen: Usual Method of Preparation and Characteristic Properties 21
9. Hydrogen—continued 24
10. Valence 27
11. Water 30
12. Chlorine 34
13. Hydrochloric Acid 36
14. Classification of Certain Compounds 39
15. Nitrogen 42
16. Ammonia 44
17. Nitric Acid 46
18. Compounds of Oxygen and Nitrogen 49
19. Carbon 52
20. Compounds of Carbon and Oxygen 55
21. Flames 58
22. Bromine 60
23. Iodine 63
24. Fluorine 66
25. Sulphur 68
26. Sulphides 70
27. Sulphur Compounds containing Oxygen 73

EXERCISE	PAGE
28. Phosphorus	75
29. Arsenic	78
30. Silicon. Boron. Review of Non-metals	80

THE MORE IMPORTANT METALS.

31. The Alkali Metals	85
32. Compounds of the Alkaline Earths	88
33. Magnesium, Zinc, Cadmium, and Mercury	91
34. Copper, Silver, and Gold	94
35. Aluminium	96
36. Tin and Lead	98
37. Chromium	100
38. Manganese	102
39. Iron, Cobalt, Nickel, and Platinum	104

SOME FAMILIAR HYDROCARBONS AND THEIR DERIVATIVES.

40. Some Hydrocarbons	109
41. Some Halogen Derivatives of the Hydrocarbons	111
42. Alcohol	113
43. Some Fatty Acids	116
44. Some Familiar Carbohydrates	118
45. A Few Aromatic or Benzene Derivatives	121

APPENDIX.

Information for Schools needing Equipment for Teaching Chemistry	125

THE NON-METALLIC ELEMENTS.

EXERCISE 1.

PRELIMINARY CHEMICAL MANIPULATION.

Select a sound cork and a sharp borer. Press the small end of the cork firmly against a block of soft wood. Place the borer in the position shown in Fig. 1, and cut a hole through the cork by twisting the borer back and forth, applying at the same time a gentle downward pressure.

A section of a properly bored cork will appear as in

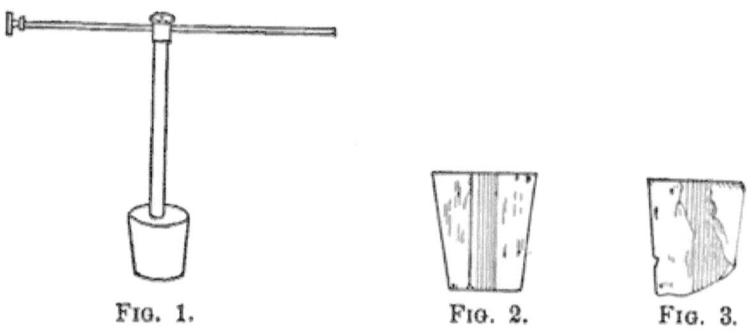

Fig. 1.   Fig. 2.   Fig. 3.

Fig. 2; when the hole is punched instead of bored the hole will appear as Fig. 3.

A hole can also be bored to advantage as follows: Proceed as just directed till the borer has cut its way about half through the cork, then remove borer, punch out the little cylinder of cork from the borer-tube, and bore from the other end of the cork. A little practice will enable you to strike the first hole, thus forming a single boring through the cork.

To fit a piece of glass tube into a cork, proceed as follows: Select a borer a trifle smaller than the tube that you wish to use. If no borer of the right size is at your disposal, select a smaller one and enlarge the hole with a round file. Now insert, by pressure and rotation, a tube which has been prepared according to directions to be found in the next paragraph.

Make a file-mark on a piece of glass tube where you wish to break it (a few inches from the end). Hold the tube as shown in Fig. 4, and break by pulling apart,

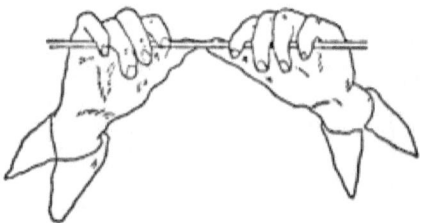

Fig. 4.

at the same time bending slightly downward. The ends of the tube will now be found sharp, and must be rounded by rotating in the flame of a burner. Take care in rounding to heat till the glass just softens, and to withdraw from flame before tube begins to melt.

For practice insert three tubes into a cork of about 2 inches diameter, as shown in Fig. 5, and submit to your teacher for approval.

Points to be observed : (1) The tubes must fit tightly. (2) The tubes must be parallel.

To fit a cork into a flask (or bottle), proceed as follows: Select a cork of such size that the diameter, at the small end, is just a little smaller than the mouth of the flask. Soften the cork by rolling between a piece of board and the work-desk as shown in Fig. 6.

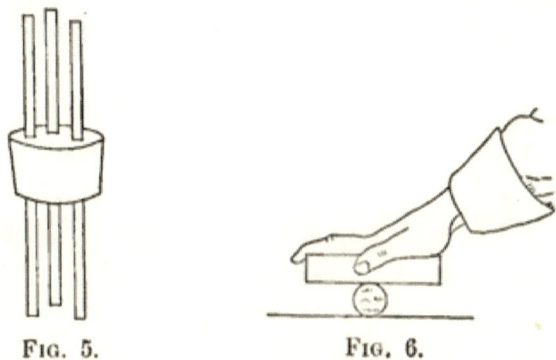

Fig. 5.   Fig. 6.

Use considerable pressure and roll for about a minute (the cork may be placed on the floor and rolled with the sole of the shoe).

For bending glass tubing the ordinary laboratory burner will not answer. The best results are obtained from a gas-burner. If the laboratory is not supplied with gas, use an ordinary kerosene lamp without the chimney. For practice, hold a piece of fairly thick-walled tube of 3 to 5 mm. in diameter outside measure ($\frac{1}{8}$ to $\frac{3}{16}$ inch) in the position shown in Fig. 7. Rotate

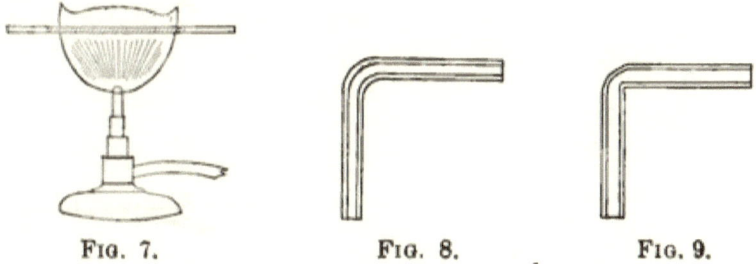

Fig. 7.   Fig. 8.   Fig. 9.

the tube, and when sufficiently hot bend to the desired angle. Now bend a tube as shown in Fig. 8, and show to your teacher for approval.

6 THE NON-METALLIC ELEMENTS.

Points to be observed: (1) There must be no marked contraction at the bend. (2) The arms must be in the same plane and at right angles.

Fig. 9 represents a piece of tube bent in a Bunsen or alcohol burner.

It is often necessary to draw out glass tubing to a smaller bore. For this purpose the tube is rotated in the flame of the burner till the glass softens. Now remove from the flame and pull the tube to the desired bore.

When you can successfully follow directions up to this point set up a wash-bottle as shown in Fig. 10.

Use a 500-cc. flask and 3 to 5 mm. diam. glass tube. The wash-bottle is to be filled with distilled water and reserved for future use.

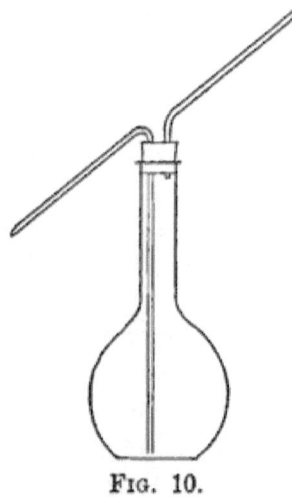

FIG. 10.

References: S. and L., pp. 395-398 and 406-497; W., pp. 384-388.

EXERCISE 2.

PHYSICAL CHANGES AND CHEMICAL CHANGES.

EXPLANATION: Physical changes do not affect the composition of substances. Chemical changes result in the formation of new substances.

In the following experiments, (a)-(f), show when chemical and when physical changes take place.

(a) Hold by means of the tongs a piece of platinum wire in the flame till it glows. Cool and examine.

(b) Try to repeat the same experiment, using magnesium ribbon instead of the platinum wire.

(c) Put a piece of tin ($\frac{1}{4}$ grm.) into a test-tube and heat till the glass begins to be red. Cool and examine.

(d) Heat a little sugar in the same way.

(e) Procure a piece of Iceland spar, weighing about 1 grm. Note its appearance and form, especially the angles. Is it soluble in water? Now lay the piece on the bottom of an inverted mortar and strike with the pestle, breaking the piece into fragments. Examine the fragments with a hand-lens, noting the form as before.

(f) Put the fragments obtained in the last experiment into an evaporating-dish and add a few drops of dilute hydrochloric acid. Warming hastens the reaction. When action ceases add a few more drops of acid, repeating the process till a further addition of a drop of the acid produces no effect. The spar should be entirely dissolved. Now place the evaporating-dish on a wire gauze supported by the ring-stand, and heat cautiously till all of the liquid is driven off. The evaporation should be carried out under the hood, or where the draft is good. Examine the residue. Is it Iceland spar? Is it soluble in water?

References: Clk., pp. 1, 2; Cly., pp. 1–3; R., pp. 1–3; S. and L., pp. 7, 8; W., pp. 1, 2.

EXERCISE 3.

Conditions under which Chemical Changes take place.

In this exercise point out the conditions—as heating, bringing into contact, submitting to the action of the light, dissolving, etc.—under which chemical changes are produced.

(*a*) Heat a little potassium chlorate ($\frac{1}{2}$ grm.) in a dry test-tube, and when the molten substance seems to boil, insert a glowing—not a flaming—splinter of wood into the test-tube.

(*b*) Place a little tin (1 grm.) in a beaker and add 1 cc. concentrated nitric acid.

(*c*) Place in the mortar a crystal of potassium chlorate as large as a grain of wheat, and in contact with it an equal amount of sulphur. Now protect the hand with a towel or glove, and grind the two substances vigorously together.

(*d*) Procure from your teacher a piece of "blueprint" paper (about 1 by 2 inches) that has not been exposed to strong light. Divide it into halves and place one in the direct sunlight for fifteen minutes. If the day is dark, leave it exposed in the window till your next period; the other half is to be kept in the dark (between the leaves of a book). After the one piece has been sufficiently exposed to the light, put both pieces in water to dissolve out any unchanged coloring matter. Compare the color of the two pieces.

(e) Grind together in a mortar equal quantities ($\frac{1}{2}$ grm. each) of sodium bicarbonate and tartaric acid. Does any change take place? Now place the mixture in a test-tube and add a little water. What do you observe?

(f) Describe the application of electricity in producing chemical change. Refer to your text-book for the answer to this question.

References: Clk., pp. 5, 6, 36, 37 ; Cly., pp. 4–7 ; R., pp. 4–7.

EXERCISE 4.

ELEMENTS, MIXTURES, AND COMPOUNDS.

In this exercise roll sulphur, not flowers of sulphur, must be used. The iron should be "iron reduced by hydrogen." Do not use iron filings.

(a) Take a few pieces of roll sulphur and carefully study its physical properties. Hold a fragment on a piece of wire in the flame of the burner and note the color of the sulphur flame, and the odor given off. Now powder the pieces in a mortar. How can you show that the powdered substance is still sulphur?

(b) Put a little of the powdered sulphur into a dry test-tube and add to it 2 or 3 cc. of carbon disulphide. (Caution: Keep carbon disulphide away from flames, as it is very inflammable.) What becomes of the sulphur in the test-tube? Pour the liquid in the tube into a watch-glass and let stand for a short time. What takes place? Explain these phenomena.

(c) Treat a little ($\frac{1}{2}$ grm.) iron with carbon disulphide just as you did the sulphur. Is the iron soluble?

Note.—Do not be confused in your conclusions by traces of iron that may be held in suspension, or by traces of sulphur that commercial carbon disulphide often holds in solution.

Dust a little iron into the flame of the burner. What do you notice? Is iron attracted by the magnet? State all the points of difference that you have observed between iron and sulphur.

You have now studied some of the properties of two elements. What is an element? See text (index).

(d) Grind together in a mortar 3 grms. of sulphur with 5 grms. of the powdered iron. Does any change take place? Examine carefully with a magnifying-glass. Can you see the iron or the sulphur? Spread out on a sheet of paper a small portion and see if you can separate the iron from the sulphur by passing the magnet just above, not quite touching the thin layer on the paper. Can you separate the iron from the sulphur by physical means?

(e) Take another portion of the material prepared in (d) and treat with carbon disulphide in a test-tube. Put your thumb over the top of the tube and shake the contents of the tube thoroughly, cork the tube and let it stand quietly for a few minutes, then pour off the liquid into a watch-glass, being careful not to disturb the solid in the bottom. Let the watch-glass stand until the liquid has evaporated. What remains? How do you know? What remains in the test-tube? How do you recognize it?

(f) What have you accomplished in this experiment? Take the rest of the material prepared in (d) and put it in a dry test-tube; heat strongly in a lamp-flame. What takes place? Break the tube and pulverize its contents in a mortar; compare it with the substances used in experiments (c), (d), and (e). Have you now a chemical compound? Why do you think so?

Write a résumé of (a) to (f) inclusive, telling what classes of substances have been used; what facts have been observed and your explanation of these facts.

References: Clk., pp. 8, 9 to Table 1; Cly., pp. 8, 9 to exercises; R., pp. 10, 12; S., pp. 8, 11; S. and L., pp. 9, 10; W., pp. 5, 6 to Sec. 8.

EXERCISE 5.

Some Chemical Terms.

This exercise contains no laboratory work. It is intended to review the preceding exercises, and to familiarize the student with certain chemical expressions that occur in the exercises that follow. The knowledge necessary to answer the questions may be derived from previous study, from explanations from the teacher, and from any work on chemistry in the school library.

Name five instances of physical change and as many instances of chemical change that you have noticed outside of the laboratory.

Can the same force under different conditions produce both physical changes and chemical changes?

Can a chemical change (as the burning of oil in the lamp) produce a physical change?

What do you understand by the terms: element; compound; mixture?

What is meant by the terms: atom; atomic weight; molecule; molecular weight?

Explain in ordinary language the expression:

$$Fe + S = FeS.$$

If the relative weights of the atoms of iron and sulphur are as 56 : 32, how much of each will you have to take to form eighty-eight pounds of iron sulphide? eighty-eight grams? eighty-eight tons?

The weight relations expressed in the problem depend on the law of the "Indestructibility of Matter." State this law. What is a natural law?

References (in addition to those given in Exercises 2, 3, and 4): Clk., pp. 10–13; Cly., pp. 10–15, 17–19; R., pp. 16–19, 71, 76–82; S., pp. 14–19; S. and L., pp. 26, 27; W., pp. 16–18.

EXERCISE 6.

OXYGEN : PREPARATION AND PHYSICAL PROPERTIES.

Preliminary Experiment.—To collect a gas not soluble in water proceed as follows : Pour water into a pneumatic trough till the water rises above the shelf. Fill a bottle with water and cover with a piece of writing-paper, taking care to exclude all air-bubbles. Now, pressing the paper against the mouth of the bottle with the fingers, place the bottle, mouth downward, in the trough and withdraw the paper below the surface of the water. Set the bottle—entirely full of water—on the shelf of the pneumatic trough. (See position of large test-tube in Fig. 11.) A tube is used to conduct the gas into the bottle. Fill the bottle with exhaled breath by blowing through a tube.

Oxygen may be collected most conveniently by displacement of water.

To prepare oxygen set up apparatus as shown in Fig. 11. Place in the small test-tube a mixture of equal weights of potassium chlorate and manganese dioxide (2 grms. each ground together in a mortar), and heat gently at first, gradually increasing the temperature till the gas comes off freely. When this stage is reached conduct the heating in such a way as to secure a gradual evolution of gas. Hold the burner in the hand and heat all parts of the mixture equally.

Collect two or three bottles of oxygen. When oxygen ceases to come off or when your bottles are full remove the delivery-tube at once from the test-tube to prevent the water from drawing back into the test-tube.

Made in this way oxygen looks cloudy on account

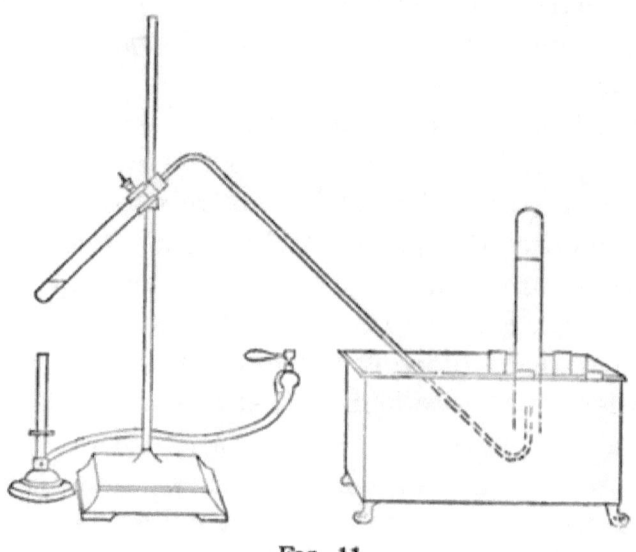

Fig. 11.

of impurities. These will nearly all dissolve out by allowing the bottles to stand for fifteen minutes in the trough. (Employ this time in writing up your notes and in making a drawing of the apparatus used.)

Now place a piece of glass plate or pasteboard below the surface of the water under the mouth of one of the bottles and invert it. Insert a splinter with a spark on the end. What takes place? This is a test for oxygen.

Try to determine whether oxygen is heavier or lighter than air in the following way: A bottle with its mouth downward, closed with a glass plate, is suddenly opened and at the same time a glowing splinter is brought up into the mouth of the bottle from below.

Another bottle with the mouth upward is uncovered and a glowing splinter inserted from above. Does the oxygen seem to fall to meet the spark or does it rise? Is the difference at all striking?

Has oxygen any odor, color, or taste?

Oxygen may also be prepared by other methods, as follows:

(a) Heat a little potassium chlorate and test as in section (a) of Exercise 3. Does this or the method last used seem to require the higher temperature?

(b) Heat a little chromic acid in a test-tube and test the escaping gas for oxygen.

(c) Prepare oxygen by heating mercuric oxide in the following way: An ignition-tube containing a little of the oxide is attached, by means of a short piece of rubber tube, to a delivery-tube. The oxide is now heated very gradually at first to avoid cracking the glass, afterwards to a high temperature. After the escape of a few bubbles (consisting of air) the oxygen evolved is collected by displacement of water. Collect in a test-tube and prove that you have obtained oxygen.

Note.—Small ignition-tubes may be made as follows: A piece of tubing long enough for two ignition-tubes is rotated in the flame till the heat gradually separates the tube into two parts of equal length. The hot parts are protected from too-rapid cooling by covering with soot from the luminous flame of a gas-burner. (A kerosene lamp with the chimney removed may be used instead of the gas burner.

EXERCISE 7.

CHEMICAL PROPERTIES OF OXYGEN. PROBLEMS.

In testing the action of gases care should be taken to prevent unnecessary diffusion; hence the bottles should be kept well covered and the substances used in the tests should be introduced through small openings. Solids may be conveniently introduced by means of a deflagrating-spoon. Such a spoon may be made of crayon and wire as shown in Fig. 12.

Oxygen unites with every other element except fluorine; hence it possesses great chemical activity. Is it chemically active with most substances at ordinary temperatures? To determine this introduce into a bottle of the gas a small piece of charcoal, and allow it to remain several minutes. Do you notice any change? Why? Repeat the same process, using a small piece of sulphur. What is the result? What do these experiments show? Now set fire to a piece of charcoal and allow it to remain in the air for a few minutes; then insert it into a jar of oxygen. Is there any difference in the manner in which it burns? What causes this difference? Repeat the process, using sulphur. What do you observe? Are the products of combustion the same when substances burn in air as when they burn in pure oxygen? Give reasons for your answer. Heat the end of a picture-wire and dip into flowers of sulphur. Set fire to the sulphur adhering to the wire and insert into a bottle containing

FIG. 12.

oxygen. Describe what takes place. What is the nature of the substance formed? Does the amount of oxygen that can be procured from 1 grm. of potassium chlorate ever vary?

Does a given amount of mercuric oxide always yield a definite amount of oxygen?

What do you understand by the "law of definite proportions"? Explain the meaning of the expressions:

$$KClO_3 = KCl + 3O.$$
$$HgO = Hg + O.$$
$$2CrO_3 = Cr_2O_3 + 3O.$$

(1) How much oxygen can you get from 1 grm. of mercuric oxide?

Calculation: From the text-book you find that oxygen and mercury have atomic weights of 16 and 200 respectively; hence mercuric oxide has a molecular weight of 216.

$$HgO = Hg + O;$$
$$216 = 200 + 16;$$

hence $\dfrac{16 \times 100}{216} = 7.41\%$ oxygen in mercuric oxide.

7.41% of 1 grm. = 0.0741 grm. of oxygen.

Second calculation:

Molecular Weight.	Atomic Weight.	Weight in grms.	Weight in grms.
HgO	: O =	Mercuric oxide :	oxygen sought.
216	: 16 =	1	: x.

$$x = \frac{16 \times 1}{216} = 0.0741 - \text{grm. oxygen.}$$

(2) How much mercury would be liberated in making 1 grm. of oxygen?

(3) How much mercuric oxide would you have to take to obtain 1 grm. of oxygen?

Solve the three following by the formula:

$$KClO_3 = KCl + 3O.$$
$$122.5 = 74.5 + 48.$$

(4) How much oxygen can you get from 1 grm. of potassium chlorate?

(5) How much potassium chloride (KCl) would be left after driving off all the oxygen from 1 grm. of potassium chlorate?

(6) How much potassium chlorate will you have to decompose in order to get 1 grm. of oxygen?

References: Clk., pp. 24-31, 72-75; Cly., pp. 83-87; R., pp. 20-36, 83-87; S., pp. 23-32; S. and L., pp. 15-18; W., pp. 19-25.

EXERCISE 8.

HYDROGEN: USUAL METHOD OF PREPARATION AND CHARACTERISTIC PROPERTIES.

Fit a wide-mouthed bottle with a good cork, through which is passed a thistle-tube and a delivery-tube. See larger bottle in Fig. 13 (the smaller bottle is not used in the first part of this exercise). Hydrogen is the lightest substance known, hence it can readily pass through parts of apparatus that are sufficiently tight to contain other gases. The cork must be well rolled and pressed very tightly into the neck of the bottle. Rubber connections may be bound with a cord or with fine wire.

Put about twenty-five grms. of granulated or sheet zinc into the bottle; replace the cork and tubes, taking care that the thistle-tube extends to within about two mm. of the bottom of the bottle. Pour through the thistle-tube dilute sulphuric acid, a few cc. at a time in order that the evolution of gas may not be too rapid. In case hydrogen is not at once generated freely, add a little copper sulphate solution through the thistle-tube. When you think that all air has been expelled from the apparatus collect several bottles of the gas by displacement of water for use in the following experiments:

(a) Thrust a glowing splinter upward into a bottle of hydrogen. Does hydrogen burn at the mouth of the bottle? Does it support combustion?

(b) Allow a bottle to stand uncovered for one to two minutes. Does it now contain hydrogen? (Test with spark.) Repeat the experiment holding the bottle with its mouth downward. Does it now contain hydrogen?

(c) Try to pour hydrogen up into a bottle in the following way: A bottle of hydrogen with its mouth downward (closed) is held beside a bottle containing air. Remove the plate from the mouth of the bottle containing hydrogen and pour the gas up into the other bottle. Does this bottle now contain hydrogen (or a mixture of hydrogen and air)?

If, now, the greater part of the zinc has dissolved, empty out the contents of the generator and recharge. When all air is expelled fill a toy balloon. Does the balloon rise? Why? Allow the balloon to rise to the ceiling. Does it fall after a time? Why?

What do you conclude from these experiments concerning the lightness of hydrogen?

Made in this way hydrogen has an odor due to impurities. Remove these by passing the gas through a solution of potassium permanganate (one part to thirty parts of water). See Fig. 13.

Collect a bottle of the pure gas and notice whether it has any odor, color, or taste.

Fill a bottle one third full of water, close tightly with the hand and place mouth downward on the shelf of the pneumatic trough; the bottle is now two thirds full of air. Fill the remaining one third with hydrogen by displacement of the water in the bottle. Remove the bottle and apply a lighted splinter, keeping the bottle well closed till the instant of ignition. This experiment shows the explosive character of a mixture of hydrogen and air. Before lighting a jet of hydrogen always collect a large test-tube of the gas by displacement of water and apply a flame at the instant of uncovering. If only a slight puff occurs and the gas

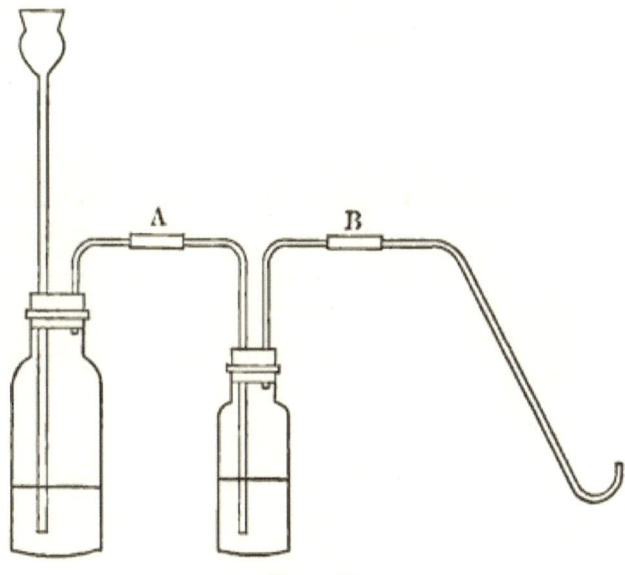

FIG. 13.

burns quietly, you can safely light the hydrogen as it issues from the generator. *Always make this test to avoid dangerous explosions.* Now remove the bottle containing the permanganate solution and connect a blowpipe with the delivery-tube of the generator at *A* (Fig. 13). Light the hydrogen as it issues from the opening in the end of the blowpipe. What is the character of the flame? Does it emit much light or much heat?

Fill a beaker with water to keep the surface cool and hold over the flame. What collects on the cool glass? What is formed when hydrogen burns?

Reserve the zinc solution in the generator for study in the next exercise.

EXERCISE 9.

HYDROGEN—CONTINUED.

The bottle used in generating hydrogen in the last exercise contains, besides undissolved zinc and impurities, zinc sulphate, either in solution or in the form of a crystalline solid. If in solution, filter and evaporate the clear filtrate on the water-bath to such a degree of concentration that crystals will gradually separate out on cooling (try two thirds former volume).

If, on the other hand, you find crystals in the bottle, dissolve them by heating the bottle and its contents in the water-bath. Filter and without concentration set aside to crystallize.

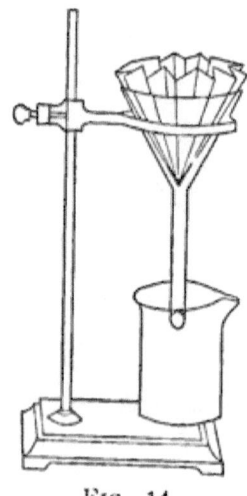

FIG. 14

The crystals consist of zinc sulphate crystallized with seven molecules of water. Show them to the instructor. (*Note on filtration.*—A circular filter-paper is folded into halves, then quarters. It is now opened out into a cone having an angle of 60°. Place it in a funnel and wet with distilled water. The paper should fit the funnel closely.)

Filtration is more rapid when a folded filter, as shown in Fig. 14, is used.

Explain the meaning of the reaction:

$$Zn + H_2SO_4 = ZnSO_4 + H_2.$$
$$65 + 98 = 161 + 2.$$

Solve the following problems according to the above reaction:

(1) How much hydrogen can you get using 100 grms. zinc?

(2) How much sulphuric acid is required to dissolve 100 grms. zinc?

(3) How much zinc and how much sulphuric acid will you require to produce a gram of hydrogen?

A liter (approximately a quart) of hydrogen weighs .0896 grm.

(4) How many liters of hydrogen will be produced in problem 3?

(5) How much zinc and how much sulphuric acid would be required to produce a liter of hydrogen?

Hydrogen can also be prepared by other methods.

(Perform experiments (a)–(d) in a test-tube and test the escaping gas with a flame.)

(a) Treat a little zinc with hydrochloric acid.

(b) Treat a little iron with dilute sulphuric acid.

(c) Perform the same experiment using hydrochloric instead of sulphuric acid.

(d) Treat aluminium ($\frac{1}{2}$ grm.) with dilute sodium hydrate solution and warm.

(e) Heat in a beaker a little water almost to boiling. Now drop in a piece of sodium as large as a grain of barley, and cover the beaker with a glass plate. What takes place? Test the solution with red litmus paper. Does it react the same as the sodium hydrate on your desk? Consult your text for symbols and write the reaction between sodium and water.

References: Clk., pp. 14-23; Cly., pp. 60-69; R., pp. 37-46; S., pp. 34-48; S. and L., pp. 34-43; W., pp. 31-38.

EXERCISE 10.

VALENCE.

You have already learned (Exercise 5) that the hydrogen atom is taken as the unit in which atomic and molecular weights are expressed. The hydrogen atom also serves as the unit in which is expressed the capacity that one atom has for combining with other atoms. This capacity is called *valence*.

1 atom of chlorine unites with 1 atom of hydrogen forming hydrochloric acid (HCl). Hence chlorine has (in this instance) a valence of 1.

1 atom of oxygen unites with 2 atoms of hydrogen forming water (H_2O). Hence oxygen has a valence of 2.

1 atom of zinc unites with 1 atom of oxygen forming zinc oxide (ZnO). Oxygen has a valence of 2, hence zinc has a valence of 2. Zinc reacts with sulphuric acid liberating 2 atoms of hydrogen for every atom of zinc that dissolves; this also shows that zinc has a valence of 2.

Lead forms with oxygen PbO and PbO_2, hence lead has a valence of 2 and of 4.

Atoms having a valence of 1 are said to be univalent; of 2, bivalent; of 3, trivalent; of 4, quadrivalent; of 5, quinquivalent.

The following list shows the usual valences of some

well-known elements. It will be seen that certain elements are given in more than one column. This shows that these elements have a varying valence.

Univalent.		Bivalent.		Trivalent.	
Bromine	Br.	Calcium	Ca.	Aluminium	Al.
Chlorine	Cl.	Carbon	C.	Antimony	Sb.
Fluorine	F.	Copper	Cu.	Arsenic	As.
Hydrogen	H.	Iron	Fe.	Boron	B.
Iodine	I.	Lead	Pb.	Iron	Fe.
Silver	Ag.	Magnesium	Mg.	Nitrogen	N.
Sodium	Na.	Oxygen	O.		
Potassium	K.	Sulphur	S.		
		Tin	Sn.		
		Zinc	Zn.		

Quadrivalent.		Quinquivalent.	
Carbon	C.	Antimony	Sb.
Lead	Pb.	Arsenic	As.
Silicon	Si.	Nitrogen	N.
Tin	Sn.	Phosphorus	P.

Write the symbols of the following-named compounds:

(1) lead iodides; (2) potassium oxide; (3) calcium sulphide; (4) magnesium iodide; (5) tin bromides; (6) hydrogen chloride; (7) aluminium chloride; (8) zinc sulphide; (9) copper bromide; (10) lead chlorides; (11) zinc oxide; (12) silicon oxide; (13) antimony chlorides; (14) carbon oxides; (15) phosphorus chlorides; (16) sodium bromide; (17) sodium sulphide; (18) sodium oxide; (19) calcium oxide; (20) arsenic oxides; (21) aluminium oxides; (22 silver bromide; (23) boron oxide.

A knowledge of valence helps us to write the symbols of compounds correctly. Example : To write the symbol silicon fluoride, we consult the above table and find that silicon has a valence of 4; fluorine of 1. Hence the symbol is SiF_4.

EXERCISE 11.

WATER.

Water has the symbol H_2O. Write the reaction that takes place when hydrogen burns. How many grams of hydrogen must you burn to produce a gram of water? How many liters of hydrogen? How many grams of oxygen will be consumed?

(a) *Distribution.*

Heat in a dry test-tube a little wood. Try the same experiment with a little bread; with sugar. What evidence have you that water is liberated in each case?

(b) *Solution in Water.*

EXPLANATION: If you treat pure salt with water in sufficient quantity, the salt disappears and you obtain a liquid that looks like pure water. The taste shows that salt is present in solution. If you add a crystal of potassium permanganate to sufficient water, the crystal disappears and you obtain a clear but highly colored solution. The color shows that the permanganate is present in the solution. On the other hand, if you shake finely powdered chalk with water you obtain a milky liquid which is not a solution since the chalk is simply suspended in the water. In the same way, oil and water, mixed, do not give a solution, as the oil is simply suspended in, or floats on the surface of, the water. Alcohol, however, dissolves in water forming a clear solution. Examples of solution: hard water, all dilute liquid reagents, clear tea and coffee. Examples of substances suspended in liquids: milk, whitewash, dirty water, muddy coffee.

Evaporate 25 cc. of "city" or well water, free from suspended matter, in a clean evaporating-dish, just to dryness. What evidence have you that the water held solid matter in solution? Repeat the experiment using distilled water. Results? Why do we use distilled water in the laboratory?

(c) *Conditions affecting Solution.*

(1) *State of Division.*—Place a crystal of copper sulphate ($\frac{1}{2}$ grm.) in a test-tube half full of water and shake for several minutes or till the crystal nearly dissolves. Repeat using the same volume of water and $\frac{1}{2}$ grm. finely pulverized copper sulphate. In which case is solution the more rapid?

(2) *Temperature.*—Add 2 grms. ammonium chloride to 5 cc. water. Shake for a few minutes. Is solution complete? Now heat to boiling. What takes place? Allow to cool and note result. What effect does heat have upon the solubility of most substances?

(d) *Saturated Solution.*

This may be prepared as follows: Put 100 cc. of water in a beaker and bring into solution as much common salt as the water will dissolve. Make use of what you have learned in the last section in regard to state of division and temperature. When no more of the salt will dissolve allow to cool to the temperature of the room. A little salt separates out, and you have a saturated solution. Filter from undissolved salt, evaporate to one half the former volume, and cool. What has separated out? What must always take place (if no chemical change occurs) when you concentrate a saturated solution? Filter the salt from the solution, dry on filter-paper, and preserve for future use.

(e) *Water of Crystallization; Water-free Crystals.*

EXPLANATION: Many substances unite with definite amounts of water in the process of crystallization. We cannot drive off this water without destroying the form of the crystal. On the other hand, many substances crystallize without water. These substances may, and commonly do, have adherent moisture on the surface or even water enclosed mechanically in the crystal, but this water has nothing to do with the structure of the crystal and may be driven off without affecting the crystal's form.

Heat a little ($\frac{1}{4}$ grm.) pure dry salt in a test-tube and note result.

Heat an equal amount of crystallized sodium sulphate in the same way. Which of these substances contains water of crystallization?

Try the same experiment with each of the following: potassium chloride, copper sulphate, and cobaltous chloride. Heat the last two substances gradually till the color just changes. Cool and add a drop of water. Note all you observe and explain fully.

(f) *Efflorescence.*

Expose to the air a crystal of sodium sulphate in a watch-glass. Examine after some time. Explain. Repeat with a crystal of sodium carbonate.

(g) *Deliquescence.*

Treat a small piece of calcium chloride and of potassium hydrate as you did the substances in (f). What do you observe in each case? (*Note.*—The time required for any marked change in (f) and (g) will depend upon the amount of moisture in the atmosphere. If the air is dry, breathe over the substances. In damp weather set in direct sunlight, in closed window.)

How can you find out whether a liquid holds a solid in solution?

Does a substance that is entirely insoluble have any taste?

EXERCISE 12.

CHLORINE.

All work with chlorine must be done under the hood. Avoid inhaling the gas.

Put 25 grms. of manganese dioxide in a (250-cc.) flask, fitted with a cork, a delivery-tube, and a thistle-tube. See Fig. 15. Add through the thistle-tube enough concentrated hydrochloric acid to cover the manganese dioxide, and shake till you are sure that all of the oxide is in contact with the acid. Warm gently on a wire gauze, taking care not to boil.

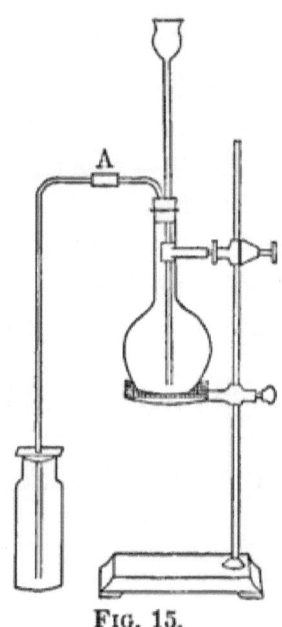

FIG. 15.

Reaction : $MnO_2 + 4HCl = MnCl_2 + 2H_2O + Cl_2$.

Collect several bottles for study by displacement of air. Try to collect one bottle by displacement of water. Which is the better way? Why?

(*a*) In one bottle place two pieces of calico (preferably red), one wet, and the other dry; a small piece each of written (ink) and printed paper, both wet; a wet green leaf or flower. Does moisture aid the bleaching action of chlorine? From which of the substances is the color discharged? Conclusion ?

(*b*) Dissolve the gas in another bottle by shaking

with 20 cc. of water. Pour half of this solution into 10 cc. of a dilute solution of some vegetable color, as indigo. Is the color discharged? Pour the rest of the chlorine-water into a solution of potassium bichromate. Does chlorine seem to have the same effect on mineral as on vegetable colors?

(c) Into another bottle of the gas sprinkle a little powdered antimony. One atom of antimony (Sb) unites with three atoms of chlorine. Write the reaction. What is the name of the product formed?

(d) Into another bottle of chlorine introduce a strip of filter-paper that has been treated with a drop of warm oil of turpentine. What takes place? Turpentine is composed of carbon and hydrogen. With which of these elements has chlorine united?

Other methods of making chlorine:

(1) Warm in a test-tube a very small amount of a mixture of salt and manganese dioxide, with a few drops of dilute sulphuric acid. Test the escaping gas with a piece of wet calico.

(2) Warm a little potassium bichromate with concentrated hydrochloric acid, and test as in (1).

(3) Put 1 cc. water in a test-tube and add the same amount of concentrated sulphuric acid. Treat a little *fresh* bleaching-powder with this acid. What is liberated?

Will chlorine burn? Will it support combustion? Is it soluble in water? How is it prepared commercially? What are its uses?

How much chlorine can be made from 60 grms. salt? How much manganese dioxide will be required?

References: Clk., pp. 111-117; Cly., pp. 70-74; R., pp. 95-100; S., pp. 92-96; S. and L., pp. 79-84; W., pp. 202-207.

EXERCISE 13.

HYDROCHLORIC ACID (HCl).

Add 30 cc. of concentrated sulphuric acid to 10 cc. of water in a beaker. While the acid is cooling set up apparatus as shown in Fig. 15, connecting at A the absorption apparatus (Fig. 16) instead of the delivery-tube. Put 20 cc. of water in the first, and several times as much in the second, bottle. The tubes should not quite reach the surface of the water. Put 25 grms. salt in the flask, replace stopper and tubes, and, when the apparatus is in order, pour the sulphuric acid through the thistle-tube and warm gently. When the gas comes off freely, disconnect at A long enough to perform the following experiments:

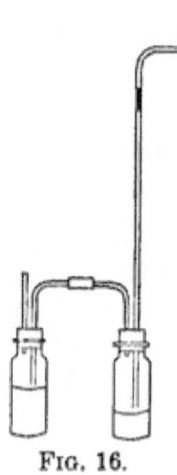

FIG. 16.

(a) Blow across the tube where the gas is escaping. What effect does the gas have on the moisture in your breath?

(b) Test the escaping gas with a lighted splinter. Does it burn? Does it support combustion?

(c) Does it affect red or blue litmus paper?

(d) Put two or three drops of ammonia on filter-paper and hold in the escaping gas. What happens?

Now connect a delivery-tube at A, as in Fig. 15, and

collect the gas in a *dry* bottle, by displacement of air. Breathe across the mouth of the bottle to determine when it is full of the gas. Cover with a glass plate and invert it in water in a pneumatic trough. Remove the plate and note what happens. Replace the plate below the surface of the water, and remove the bottle and contents. Test the liquid with red and with blue litmus paper. Place a drop on the tongue. Taste? Connect the flask again at A with the absorption apparatus and continue heating till no more gas is given off. Allow the flask to cool. It contains crystals of acid sodium sulphate ($HNaSO_4$). Now complete the reaction:

$$NaCl + H_2SO_4 = \ldots$$

Test a little of the liquid in the first bottle, in separate test-tubes, with each of the following: iron filings, zinc, and manganese dioxide (warm). What takes place in each case? Reactions?

Test the liquid with red and with blue litmus paper. To a little of the same solution (first bottle) add in separate tests a few cc. of the following solutions: silver nitrate ($AgNO_3$), mercurous nitrate ($HgNO_3$), and lead nitrate ($Pb(NO_3)_2$). Write reactions noting that 1 molecule of lead nitrate reacts with 2 molecules of hydrochloric acid, while each of the other nitrates reacts with 1.

Allow the lead chloride to settle and pour off the liquid as completely as possible without disturbing the solid. Add a few drops of water and heat to boiling, continuing the process till solution is just effected. Allow to cool. What do you observe? What do you conclude as to relative solubility of lead chloride in hot and in cold water?

Other methods of preparing hydrochloric acid:

(1) Treat a little ammonium chloride in a test-tube with concentrated sulphuric acid and test the escaping gas.

(2) Treat a little potassium chloride in the same way.

(3) Add a little concentrated sulphuric acid to a little concentrated hydrochloric acid from the bottle on your desk. Is the hydrochloric acid displaced from its solution (concentrated hydrochloric acid means a strong solution of the gas in water) by sulphuric acid?

References: Clk., pp. 118–121; Cly., pp. 76–80; R., pp. 103–108; S., pp. 97–98; S. and L., pp. 68–71; W., pp. 110–114.

EXERCISE 14.

CLASSIFICATION OF CERTAIN COMPOUNDS.

For convenience many compounds (by no means all) can be classified as acids, bases, and salts, or as simple derivatives of these.

(*a*) ACIDS.—Write the names and formulas of the three acids that you have used in these exercises. Find the names and formulas of five other acids in your textbook. What element is present in all acids? Put a drop of sulphuric acid in 25 cc. of distilled water, moisten a glass rod with this diluted acid, and taste. Repeat the same experiment, using nitric acid instead of sulphuric acid. Test each with red and with blue litmus paper. Results? (*Note.*—Use very small pieces of litmus paper, a fresh piece for each test.) Now tell all you can about the characteristics of acids. Does your description apply to *all* (as very weak or insoluble) acids? To answer this, test as above a little of a stearine candle (which consists largely of stearic acid). Results?

(*b*) BASES.—Try the action of each of the following solutions with litmus paper: sodium hydrate, potassium hydrate, ammonium hydrate ("ammonia"), barium hydrate, and calcium hydrate ("lime-water"). What do you observe?

Put a little solid magnesium oxide on red litmus paper and moisten; allow to stand several minutes. Is the litmus changed? Taste a drop of very dilute solution of sodium hydrate; of calcium hydrate. What have you learned about bases? Do these characteristics apply to insoluble or very weak bases or basic oxides? To answer, test the action of moistened zinc oxide or copper oxide on red litmus paper.

(c) SALTS.—Dissolve a little pure sodium chloride in distilled water in a clean test-tube and test the solution with litmus paper of both colors. Is either affected? Repeat with potassium chloride; magnesium sulphate; sodium sulphate; potassium nitrate.

Dissolve 2 grms. sodium hydrate in 5 cc. water in an evaporating-dish by the aid of gentle heat. Add dilute hydrochloric acid 1 cc. at a time till the solution is just neutral. After each addition, test in the following way: A glass rod, drawn to a point, is dipped into the solution; the point is placed on blue litmus paper. As soon as the acid is in excess the spot where the rod touches becomes red. Stir to be sure that the solution is well mixed and test again. If still red, you have acid in excess. Now add cautiously sodium hydrate from the reagent bottle, drop by drop, stirring well and testing after each addition, till the solution changes neither red nor blue litmus. If you accidentally add a trifle too much sodium hydrate, neutralize with a very dilute acid solution. Filter if the solution is not perfectly clear and evaporate to about one half the former volume. Now test again with litmus to make sure that you made no error when you decided that the solution was neutral. If not perfectly neutral, neutralize. Now place the dish on a wire gauze and evaporate to dryness over the open flame. What is the substance? Reaction by which it is made? Weigh it. If you used exactly 2 grms. of sodium hydrate, how much salt ought you to get by the above reaction?

Do all salts give a neutral reaction with litmus? Test (in solution) sodium carbonate, potassium carbonate, copper sulphate, alum, and iron (ferric) chloride. Do weak acids perfectly neutralize the strong basic properties of sodium and potassium? Does weakly basic iron counteract the acid properties of chlorine in ferric chloride?

References: Clk., p. 61 ; Cly., pp. 43-47; R., pp. 116-132; S., pp. 75-79 ; S. and L., pp. 58-62 ; W., pp. 89-96.

EXERCISE 15.

NITROGEN.

Nearly pure nitrogen can be obtained from the air by burning phosphorus in a closed vessel. *Phosphorus should be kept and cut under water. Handle with pincers, and throw all unconsumed pieces in the slop-jar. After using, heat the deflagrating-spoon to ignite any pieces that may adhere to it.*

Make a deflagrating-spoon with some copper wire and a piece of crayon hollowed at one end. Bend in the shape shown in Fig. 17.

Place a piece of phosphorus as large as a pea in the cup of the spoon. Light the phosphorus, insert in an inverted bottle, and immediately immerse the mouth of the bottle about an inch below the surface of the water in the pneumatic trough. A little air is first driven out. Why is this? Note carefully the burning. The white fumes are phosphorus pentoxide (P_2O_5). In a short time the cloud disappears. Where does it go? Why does the water rise in the bottle? What proportion of the air remains? What has become of the oxygen? Allow the bottle to stand in the water for fifteen minutes, then slip a glass plate under the bottle below the surface of the water and invert the bottle and its contents.

FIG. 17.

Has nitrogen any color, odor, or taste? Will it burn? Will it support combustion? Test with a glowing splinter; also with burning sulphur. Is nitrogen chemically active?

Consult your text if necessary in answering the following questions:

What purpose does nitrogen serve in the air? How can it be shown that air is a mixture and not a compound? What part of the air is nitrogen, by volume and by weight? Is nitrogen lighter or heavier than air? than oxygen?

References: Clk., pp. 49-54; Cly., pp. 115, 116, 124-126; R., pp. 126-128; S., pp. 50, 51, 82-85; S. and L., pp. 10-14, 18-20; W., pp. 26-30, 185-189.

EXERCISE 16.

AMMONIA.

Set up apparatus as in Exercise 13, without the thistle-tube. See Figs. 15 and 16. Use an asbestos plate instead of the wire gauze. Place 10 grms. of quicklime (free from air-slaked lime and stone) in the flask, add 10 cc. water and warm gently. Allow to cool and add 15 grms. of ammonium chloride. Mix by cautiously shaking.

Set up and fill the absorption apparatus exactly as in Exercise 13. Now heat, and when the gas comes off freely disconnect at A long enough to perform the following experiments:

(a) Test the escaping gas with a glowing splinter. Does it burn? Does it support combustion?

(b) Test it with red and with blue litmus paper. Results?

(c) Collect a bottle of ammonia just as you collected hydrochloric acid, by displacement of air—holding, however, the mouth of the bottle downward. Let the tube reach up almost to the bottom of the bottle. To tell when full, test with a piece of moistened red litmus paper held near the mouth of the bottle. Test the solubility as you did in the case of hydrochloric acid.

(d) Is ammonia lighter or heavier than air? Does the gas seem to diffuse upward or downward most rapidly as it escapes at A? Test with litmus.

Now connect the absorption apparatus and continue to heat as long as gas comes off freely. Too long heating is liable to crack the flask.

The first bottle should contain a strong solution of ammonia in water. Since this solution acts very much like potassium hydrate, we assume that ammonium hydrate has been formed according to the reaction: $NH_3 + H_2O = NH_4OH$.

Divide the solution into two parts. Neutralize one with hydrochloric acid and evaporate to dryness over a low flame. What remains? Reaction by which it was formed? Heat a little in a dry test-tube. What collects on the sides of the test-tube? This process is called *sublimation*. Treat a little of the sublimate in a test-tube with sodium hydrate solution and warm. Odor given off?

Neutralize the other half of the ammonium hydrate solution with dilute sulphuric acid. Reaction? Evaporate and test the residue with sodium hydrate. What is given off? How did you recognize it? Complete the reaction: $2NaOH + (NH_4)_2SO_4 = \ldots$

Note.—In case your ammonium hydrate solution proves too weak to give the salts in sufficient quantity, use the ammonia solution on your desk for neutralization with the acids.

Has ammonia any odor or color? Has it neutral, acid, or basic properties? Compare it with hydrogen and with nitrogen. Does it resemble the elements of which it is composed? What is the source of the ammonia of commerce? What are some of the uses of ammonia?

References: Clk., pp. 53–57; Cly., pp. 117, 118; R., pp. 133–141; S., pp. 52–58; S. and L., pp. 62–67; W., pp. 133–138.

EXERCISE 17.

NITRIC ACID.

The apparatus used in preparing nitric acid is shown in Fig. 18. The retort should be small (100 to 200 cc.). Place in the retort 15 grms. sodium nitrate and 10 cc. concentrated sulphuric acid. Fill the beaker containing the test-tube nearly full of cold water. Now heat the gauze gently, as long as a liquid collects in the test-tube used as a receiver; then remove the burner and allow the retort to cool. Note the color and odor of the liquid collected. Is it the same compound as the concentrated nitric acid on your desk? Try the action of a few drops of each (diluted with a little water) on copper. Place a small piece of finger-nail in a drop of the undiluted acid and note the color after washing with water. The same result is produced on many animal substances.

FIG. 18.

Nitric acid is colorless. The reddish-brown gas seen in the retort and receiver is nitrogen dioxide. The nitric acid may be colored by some of this substance in solution. When the retort is cool, pour in a little water and heat carefully, adding more water, 1 cc. at a time, till the substance ($NaHSO_4$) is dissolved. Filter the solution thus obtained. If you have not added too much water, crystals will separate out on cooling. If none appear, concentrate by evaporation (one half former volume) and cool again. When a good crop of crystals is obtained, pour off the liquid, redissolve the crystals in as small a quantity as possible of distilled water, and crystallize as before. Pour off the liquid and dry crystals on filter-paper. Dissolve one or two in distilled water. Is the solution acid, basic, or neutral? Label the crystals sodium bisulphate and hand them to your instructor. Complete the reaction: $NaNO_3 + H_2SO_4 = NaHSO_4 + \ldots$

Sodium bisulphate is an acid salt. Compare its symbol with the symbol of sodium sulphate and of sulphuric acid.

Now explain what is meant by the term "acid salt."

Action of Nitric Acid on Metals.

(a) How did you make hydrogen from sulphuric acid? From hydrochloric acid? Can you make hydrogen by the action of nitric acid on a metal? Try with zinc; with iron. Test the escaping gas in each case. Can you detect hydrogen? Nitric acid is a good oxidizing agent. It oxidizes the hydrogen to water.

(b) Place a little tin in a test-tube and just cover with concentrated nitric acid. Heat gently. If action does not begin at once, add a few drops of water and warm again. Examine the substance left. It is a compound consisting mostly of tin and oxygen, the latter derived from the nitric acid.

What are some of the most important salts of nitric acid? How are they obtained? What use is made of nitric acid in the manufacture of high explosives?

PROBLEMS.

(1) How much nitric acid can be made from 150 grms. of sodium nitrate?

(2) From the same quantity of potassium nitrate?

(3) 300 grms. nitric acid are required; how much sodium nitrate must be used?

(4) How much sodium bisulphate will be formed in Problem 3?

References: Clk., pp. 58–60; Cly., pp. 119, 120; R., pp. 143–148; S., pp. 67–70; S. and L., pp. 55–57; W., pp. 116–122.

EXERCISE 18.

COMPOUNDS OF OXYGEN AND NITROGEN.

(a) *Nitrous Oxide* (N_2O).

Set up apparatus as shown in Fig. 11. Heat in the test-tube 5 grms. of ammonium nitrate till it seems to boil. Drive off the gas at as low a temperature as possible and collect in bottles by displacement of water. Allow the bottles to stand in water for fifteen minutes before examining; then test color, odor, and taste. Will the gas burn? Will it support combustion? Is it soluble in water to any marked extent?

Complete the reaction NH_4NO_3 (heated) = ...

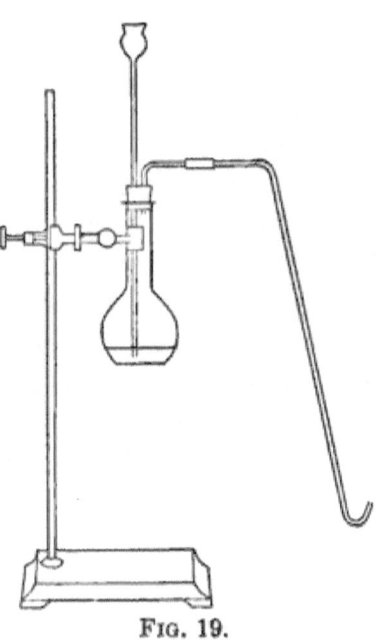

FIG. 19.

(b) *Nitric Oxide* (NO).

Use apparatus shown in Fig. 19. Put 15 grms. of copper in the flask and add dilute nitric acid through the thistle-tube. Collect the gas by displacement of water. Add the acid a little at a time so that the gas may come off gradually. Warm a little now and then, regulating the supply of acid and the temperature so as to produce a regular evolution of gas. If the gas begins to come off too rapidly, pour in cold water through the thistle-tube.

Reaction : $3Cu + 8HNO_3 = 3Cu(NO_3)_2 + 4H_2O + 2NO$.

Test the properties of nitric oxide. Will it burn? Will it support combustion?

(c) Nitrogen Peroxide (NO_2).

Turn upwards one of the bottles containing the colorless nitric oxide just obtained and uncover. The gas absorbs oxygen from the air and becomes the reddish-brown nitrogen peroxide. Write the reaction. Does nitrogen peroxide burn? Does it support combustion? Is it soluble in water? Add a little water to a bottle of the gas, cover with the hand, and shake well. Does the color disappear? Now explain the colored appearance of the gas in the flask used in preparing nitric oxide, and the fact that a colorless gas was collected by displacement of water.

Which of the compounds studied in this exercise most resembles oxygen? How can you explain the fact that the one that contains the smallest per cent of oxygen is the only one that you found to be a supporter of combustion?

(e) The Law of Multiple Proportions

is well illustrated by the compounds studied in (a), (b), and (c). We will assume that nothing is known about atoms, molecules, and symbols. All that we need to know is that we have three distinct compounds (not mixtures), each composed of oxygen and nitrogen. It can be shown that heated copper has the property of absorbing all the oxygen in any of

COMPOUNDS OF OXYGEN AND NITROGEN. 51

them, leaving behind all the nitrogen. If we make such an analysis in each case we shall obtain the following results :

in 11 grms. nitrous oxide, } 7 grms. nitrogen, 4 grms. oxygen.

in 15 grms. nitric oxide, } 7 grms. nitrogen, 8 grms. oxygen.

in 23 grms. nitrogen peroxide, } 7 grms. nitrogen, 16 grms. oxygen.

How can the simple ratio $4:8:16$ $(1:2:4)$ be explained? Answer: By assuming that compounds are made up of simple units or atoms. Hence the simple ratio of weights is to be explained by assuming a simple ratio in the numbers of the atoms in the compounds.

That the weight relations just considered are in harmony with the molecular weights as derived from the symbols given in the exercise can be readily seen from the following :

nitrous oxide, $\quad 7: 4 = 14 + 14 : 16 \quad \ldots \quad N_2O.$
nitric oxide, $\quad 7: 8 = \quad\quad 14 : 16 \quad \ldots \quad NO.$
nitrogen peroxide, $7: 16 = \quad\quad 14 : 16 + 16 \quad . \quad NO_2.$

State the law of multiple proportions.

EXERCISE 19.

CARBON.

Heat in a porcelain or metal crucible one piece each of hard and of soft wood covered with sand. Light the gas above the sand. Does the charcoal from the different woods differ in physical properties? Put a little powdered charcoal on platinum foil and heat strongly. Does it burn? Try the same experiment with graphite (another form of carbon) from your pencil. Is carbon active at ordinary temperatures? Is it attacked by acids or alkalies? Try with those on your desk. Carbon in the form of "lamp-black" is used in preparing printer's ink. Why is printer's ink as well as India ink so permanent?

Name and describe three allotropic forms of carbon.

(a) Charcoal as an Absorbent.

To a test-tube one third full of water add indigo solution till the water is distinctly blue. Now add 1 grm. of animal charcoal and heat, constantly shaking for several minutes. Filter, and note whether the blue color has disappeared.

Repeat the experiment, using litmus solution instead of indigo. Results?

Dilute hydrogen sulphide water till the odor is only just distinctly noticeable. Shake with animal charcoal and note odor before and after shaking.

Results? What do you conclude from these experiments concerning the absorbent power of carbon in the form of animal charcoal?

(b) *Carbon as a Reducing Agent.*

(1) Mix by grinding in a mortar 3 grms. of powdered copper oxide with 0.2 grm. of wood charcoal. Put in a small porcelain crucible and cover with a very thin layer of pulverized charcoal, then place the crucible cover on the crucible to protect the mixture from the action of the air. Place the crucible and its contents on a wire or pipe triangle, and heat strongly for five minutes with Bunsen or gasoline burner. See Fig. 20. (Few alcohol-lamps give sufficient heat for this reaction.) Cool and examine the contents of crucible. Has the copper oxide been reduced to copper?

Fig. 20.

Try the action of some of the reddish substance with nitric acid. Complete the reaction

$$2CuO + C = CO_2 + \ldots$$

(2) Heat a little lead oxide (PbO) on charcoal before the blowpipe. (The use of the blowpipe is fully described in most works on chemistry. See index in text or reference-books.) Is the lead oxide reduced by the heated charcoal? Complete the reaction $2PbO + C = \ldots$

(c) *Test for Carbon.*

Carbon is most readily detected by burning it to carbon dioxide. Make several bottles of oxygen as in Exercise 6. Cover with greased glass plates and reserve for the following experiments:

(1) Fasten a piece of charcoal to a piece of wire, heat to glowing, and burn in a bottle of oxygen. As soon as the charcoal ceases to glow, add 5 cc. of clear lime-water and shake. What do you notice?

Reaction: $Ca(OH)_2 + CO_2 = CaCO_3 + H_2O$.

$CaCO_3$ is insoluble in water, hence the liquid becomes turbid.

(2) Repeat the experiment, using a piece of candle instead of charcoal. Does the candle contain carbon? Why do you think so? How would it be possible to prove that the diamond is carbon?

References: Clk., pp. 74-79; Cly., pp. 147-152; R., pp. 156-167; S., pp. 125-131; S. and L., pp. 166-176; W., pp. 44-52.

EXERCISE 20.

COMPOUNDS OF CARBON AND OXYGEN.

(a) *Carbon Dioxide.*

Set up apparatus consisting of a large test-tube carrying a thistle-tube and a delivery-tube. See Fig. 21. Put a little solid sodium carbonate into the test-tube and add dilute hydrochloric acid through the thistle-tube. Pass the escaping gas into lime-water.

Now try in separate tests the action of sodium carbonate with dilute sulphuric, nitric, and acetic acids. Does a gas escape which renders lime water turbid in each case?

FIG. 21.

Try the action of marble (calcium carbonate) with each of the acids mentioned above. Is the same gas liberated as before?

Complete the reactions:

$$Na_2CO_3 + 2HCl = H_2O + \ldots$$
$$Na_2CO_3 + H_2SO_4 = H_2O + \ldots$$
$$Na_2CO_3 + 2HNO_3 = H_2O + \ldots$$

Note that calcium replaces two atoms of hydrogen and complete the following:

$$CaCO_3 + 2HCl = \ldots$$
$$CaCO_3 + H_2SO_4 = \ldots$$
$$CaCO_3 + 2HNO_3 = \ldots$$

Prepare carbon dioxide for study by the action of hydrochloric acid on marble. Use apparatus shown in Fig. 15, and collect the gas by displacement of air. Will the gas burn or support combustion? Test with a lighted splinter. Is it lighter or heavier than air? Pour it as you would water on a lighted splinter. Result? Pour the gas in one bottle into another, and test for carbon dioxide in the second bottle with a burning splinter. Is carbon dioxide soluble in water? Test as follows: Collect a bottle of the gas by displacement of air. Now remove delivery-tube and insert a thistle-tube through the hole in the pasteboard cover. Fill the bottle one third full of cold water through the thistle-tube. Remove tube and cover, close tightly with the hand and shake vigorously for a minute or two. After putting the mouth of the bottle under water, remove the hand and note result.

Dissolve 2 grms. of solid sodium hydrate in 10 cc. distilled water and allow to cool. Place in a test-tube and pass in carbon dioxide rapidly. Does the solution become warm? After the gas is no longer absorbed test the solution with acids. Does it behave like a solution of sodium carbonate?

Complete the reaction:

$$2NaOH + CO_2 = \ldots$$

Why do you get a precipitate with calcium hydrate and not with sodium hydrate? Compare the solubility of the two carbonates.

Pass your breath through a solution of calcium hydrate in a test-tube. What does this show? Allow a little of the hydrate solution to stand exposed to the air for several hours. What evidence have you that the air contains carbon dioxide?

What use is made of carbon dioxide by growing plants? (See text.)

(b) *Carbon Monoxide.*

Connect to the apparatus used in generating carbon dioxide (Fig. 15) a wash-bottle such as you used in purifying hydrogen (Fig. 14). Put into the flask 10 grms. oxalic acid, and add 30 cc. concentrated sulphuric acid through the thistle-tube. Fill the wash-bottle half full of sodium hydrate solution. Now heat gently at first, afterwards more strongly, and, when air is expelled, collect gas by displacement of water.

Reaction: $C_2H_2O_4 = CO_2 + CO + H_2O$.

The sulphuric acid serves to remove water from the oxalic acid. Why is the wash-bottle containing sodium hydrate used?

Give physical properties of carbon monoxide. Compare it in all respects with carbon dioxide. Explain how the two oxides of carbon illustrate the law of multiple proportions.

References: Clk., pp. 90–98; Cly., pp. 153–156; R., pp. 168–179; S., pp. 136–145; S. and L., pp. 176–183; W., pp. 165–178.

EXERCISE 21.

FLAMES.

(a) *The Nature of Flames.*

Heat a piece of charcoal till it glows. Here is combustion, but no flame. Did you get a flame when glowing charcoal was plunged into oxygen, or only a more intense glowing? Now light a candle. Have you here a flame? After it is burning well, blow it out suddenly, and hold a lighted splinter a little above the wick. Can you relight the candle without applying the light to the wick?

Place 1 grm. of stearin (from a candle) in a small crucible. (See Fig. 20.) Heat strongly. Are gases evolved? Can you light them above the crucible?

Repeat the experiment, using a few drops of alcohol. Results? Fill a test-tube one-third full of bits of wood and heat. Are combustible gases given off? Do you think that any substance that burns with a flame will, on heating, give off combustible gases? If so, what is a flame?

(b) *Kindling-point.*

Do all substances begin to burn at the same temperature? Test with the following: 1 cc. of carbon disulphide, in a small beaker; a match; and a candle. Warm a glass rod and hold in the vapor of the carbon disulphide, then immediately touch each of the other two. Repeat the experiment with the rod a little hotter each time, till one of the substances is kindled. Now see which of the remaining two you can light with a hot rod. Does each substance seem to have its own distinct kindling-point?

Bring down a piece of wire gauze on the flame of a burner. Does the flame pass through the gauze? Apply a light above the gauze. What takes place? Explain fully the principle of the safety-lamp.

(c) *Light of Flames.*

Could you read well by the light of an alcohol or a hydrogen-flame? Hold a clean evaporating-dish in burning alcohol. Is any soot (carbon) deposited? Try the same experiment with a candle-flame. To what is the light of an ordinary flame principally due?

Alcohol is a little over 50% carbon. Oil of turpentine is nearly 90% carbon. Which do you think will burn with the more luminous (and sooty) flame? Light a very little of each, and see if your reasoning is correct.

References: Clk., pp. 84-89; Cly., pp. 159-171; R., pp. 180-189; S., pp. 27-29; S. and L., pp. 183-193; W., pp. 53-62.

EXERCISE 22.

BROMINE.

Work with bromine must be performed under the hood or where the draft is good. Avoid inhaling the fumes.

(*a*) *Bromine-vapor.*

Place in a large test-tube 2 or 3 crystals of potassium bromide with an equal quantity of manganese dioxide. Add a few drops of concentrated sulphuric acid and warm. Note color of vapor. Will it bleach? Try with moistened calico. How does the action compare with that of chlorine?

(*b*) *Bromine* (*Liquid*) *and Bromine-water.*

Put a pulverized mixture consisting of 3 grms. potassium bromide and an equal weight of manganese dioxide into retort of apparatus shown in Fig. 18. Pour 10 cc. concentrated sulphuric acid into 50 cc. of water, and when the liquid is cool introduce it by means of a funnel into the retort. Shake to be sure that the acid is in contact with all parts of the mixture and then heat gently. Put a little water into the test-tube used as a receiver, and let the end of the retort dip below the surface of the water. This is to prevent evaporation. Now continue to heat till brown fumes no longer come off.

. The heavy brown liquid in the test-tube is bromine. Compare it with chlorine, giving the physical properties of each. The reaction representing the liberation of bromine is exactly analogous to the reaction that you learned in preparing chlorine. Write the reaction for bromine.

How do you account for the color of the water above the bromine? Is bromine soluble in water?

Bromine is very soluble in carbon disulphide. To a few cc. of bromine-water (free from liquid bromine) in a test-tube add 1 drop of carbon disulphide, close the test-tube with the thumb and shake vigorously. Allow to settle. Which now contains the bromine in solution, the water or the carbon disulphide? Note the color of each. Do you think that you could prove the presence of bromine in solution by this test? Now try the same test with a solution of potassium bromide. Can we use this test for compounds containing bromine or only for bromine itself (free bromine)?

To a few cc. of a dilute solution of potassium bromide add a little chlorine-water. Does the solution become colored? Now shake with a drop of carbon disulphide. Has bromine been liberated by the chlorine? Which has the greater affinity for potassium, bromine or chlorine? Complete the reaction: $KBr + Cl = \ldots$

(c) *Hydrobromic Acid and some of its Salts.*

Put a crystal of potassium bromide in a test-tube, add a little dilute sulphuric acid and heat. Test the escaping gas with litmus paper. Write the reaction representing the liberation of hydrobromic acid. See corresponding reaction for hydrochloric acid.

In making the following bromides we may use either hydrobromic acid or a soluble bromide.

$$AgNO_3 + HBr = AgBr + HNO_3.$$
$$AgNO_3 + KBr = AgBr + KNO_3.$$

Add a few drops of potassium bromide solution, in separate tests, to solutions of silver nitrate, mercurous nitrate, and lead acetate. Describe results and write reactions.

(*d*) PROBLEMS.

1. How much bromine can be obtained from 60 grms. potassium bromide?

2. How much potassium bromide must be taken to obtain 60 grms. of bromine?

3. How much potassium bisulphate will be formed in problem 2?

References: Clk., p. 128; Cly., p. 75; R., pp. 217–220; S., pp. 108–114; S. and L., pp. 87–90; W., pp. 208–211.

EXERCISE 23.

IODINE.

Grind together in a mortar 2 grms. of potassium iodide and 4 grms. of manganese dioxide. Place the mixture in a beaker of 300 to 400 cc. capacity and add drop by drop about 5 cc. of concentrated sulphuric acid. Select an evaporating-dish whose diameter is a little larger than that of the beaker. Fill the evaporating-dish three fourths full of water, see that the under surface is perfectly clean and dry, and place on the beaker. Now set beaker and contents, covered as just directed, on a wire gauze supported by the ring-stand. Heat cautiously and note the color of the iodine-vapor. Conduct the heating in such a way that the iodine-vapor just fills the beaker. (Too rapid heating is to be avoided. Half an hour will be required to obtain good results.) If the iodine collects on the sides of the beaker, vaporize it by cautious heating. Continue the heating till nearly all the iodine has collected on the cool surface of the evaporating-dish. Then remove the dish, throw out the water without wetting the iodine, and scrape the iodine off the dish. Now refill the dish with cool water, and see if you can sublime any more iodine against it. If so, add it to the portion first obtained.

The liberation of iodine is analogous to the liberation of chlorine or bromine. Write the reaction for iodine. Describe in detail everything that you have noticed in connection with the preparation of iodine.

The most delicate test for iodine is a solution of starch paste. Prepare this as follows: To 1 grm. starch add 10 cc. water, mix well and pour into 200 cc. boiling water. Stir well. Allow to cool and settle, and pour off the clear liquid for use when required. When the solution is cooling test the solubility of iodine in the following liquids: water, alcohol, carbon disulphide, and a solution of potassium iodide. Use a test-tube full half of (hot) water and 4 or 5 drops of each of the other solvents. In which do you think that iodine is the most soluble?

Now make the "starch test" as follows: Add a few cc. of the starch solution to 100 cc. of water and then add the water solution of iodine till a deep blue appears. If you do not succeed in this way, dilute the solution of iodine in potassium iodide with a large excess of water and add a drop of this dilute solution. The success of the experiment depends on the relative amounts of starch, water, and iodine.

To another portion of the water solution or to a very dilute potassium iodide solution of iodine add a drop of carbon disulphide and shake well. What takes place? Compare the deportment of bromine with this reagent.

Dissolve a minute amount of potassium iodide in a test-tube half full of water. Does the solution react with starch? Now add a drop or two of chlorine-water and note result. Shake a little of potassium iodide solution with a drop of carbon disulphide. Does any change of color occur? Now add a drop or two of chlorine-water and shake again. Result? Explain fully the principles involved.

IODINE.

Treat silver nitrate, mercurous nitrate, and lead nitrate with a solution of potassium iodide. Describe results and write reactions. Compare with the action of potassium bromide.

PROBLEMS.

(1) How much potassium iodide must you take to obtain 100 grms. of iodine?

(2) How much manganese dioxide would be required in problem (1)?

References: Clk., pp. 129–131; Cly., p. 75; R., pp. 220–222; S., pp. 115–118; S. and L., pp. 90–93; W., pp. 211–215.

EXERCISE 24.

FLUORINE.

Give a brief description of the properties of fluorine using your text-book or lecture-notes as your source of information. Why do we not prepare fluorine for study in the laboratory?

Hydrofluoric Acid.

This compound attacks glass, hence we cannot well study it very fully. It is liberated from its salts by the action of sulphuric acid. Write the reaction that takes place between calcium fluoride (CaF_2) and sulphuric acid. The action of hydrofluoric acid on glass may be shown in the following way: Cover a piece of window-glass with a coating of beeswax or paraffine somewhat thicker than the paper of this book. This is best done by carefully warming the glass and rubbing the wax over its surface. The coating should be on both sides of the glass and as uniform as possible. Engrave some design in the wax by means of a pointed instrument. Be sure that the point of the instrument entirely removes the wax and that the lines are not too fine. Put 3 grms. calcium fluoride in a leaden dish and add enough conc. sulphuric acid to form a thin paste. Cover the dish with the plate that you have prepared with the engraved side down. Avoid inhaling the fumes. Allow the dish thus covered to stand for several hours, or till your next laboratory period, under the hood. Next remove the plate and at once wash the contents of the dish into the waste-jar. It is sometimes troublesome to remove the wax. This can be done without cracking the glass by putting in water, heating to boiling, continuing the boiling for a few minutes, then cooling till the hand can be placed in the water without scalding. The plate may now be removed and rubbed with a cloth moistened with alcohol. Show the etched plate to your instructor for approval. The reaction between the acid and the glass is as follows:

$$SiO_2 + 4HF = SiF_4 + 2H_2O.$$

We have, for convenience, studied the halogens (fluorine, chlorine, bromine and iodine) in an illogical order. Now make a thorough review of these compounds. Show in what respects they resemble each other. Make a list of their atomic weights and show that their properties are related to their atomic weights.

References: Clk., pp. 111-113; Cly., pp. 75, 76; R., pp. 223-225; S., pp. 122-124; S. and L., pp. 77-79, 94-95; W., p. 215.

EXERCISE 25.

SULPHUR.

(a) *Rhombic Sulphur.*

Dissolve in a small beaker 1 grm. of roll sulphur in carbon disulphide, using as small a quantity as will effect solution. Set the beaker under the hood, *away from flames*, and allow the solvent to evaporate. Note the form of the crystals.

(b) *Monoclinic Sulphur.*

Fill a porcelain crucible with roll sulphur and melt at a low temperature. Add more sulphur as it melts down, continuing the heating at a *low* temperature, till the crucible is full of the molten sulphur. Now allow to cool until a crystalline crust begins to form on the surface. Make an opening and pour out the remaining molten sulphur. Note the kinds of crystals in the cavity. Allow them to stand several days. Do you notice any change in general appearance or form?

(Write out a brief description of each of the six systems of crystals. See text or reference-books.)

(c) *Plastic Sulphur.*

Put about 5 grms. sulphur in a dry test-tube, held with a clamp, and heat gradually. Note that you first obtain a thin amber-colored liquid which gradually becomes dark and viscous as the temperature rises.

This soon becomes thinner, and finally the tube is filled with sulphur-vapor. Pour the molten sulphur into a beaker containing water. Examine and describe the sulphur in the beaker. Allow it to stand till your next laboratory period. Is it still plastic?

Now compare the three allotropic forms of sulphur. Which form is the most stable? What substance have you studied all of whose allotropic forms are stable?

Where is the greater part of the sulphur of commerce obtained? How is it freed from impurities?

References: Clk., pp. 132-137; Cly., pp. 101-103; R., pp. 226-230; S , pp. 157-160; S. and L., pp. 99-105; W., pp. 238-243, 108.

EXERCISE 26.

SULPHIDES.

(a) *Hydrogen Sulphide.*

All work with hydrogen sulphide must be done under the hood or where the draft is good. Put all residues in waste-jar under the hood or out of doors. Avoid inhaling the gas.

Set up apparatus exactly as in Fig. 13, except the delivery-tube. Put 20 grms. ferrous sulphide in the generator and 50 cc. of water in the small bottle. Attach to this bottle a delivery-tube such as you have used in collecting gases by displacement of air. When the apparatus is in order, pour dilute sulphuric acid into the generator through the thistle-tube. After all air is expelled, collect a bottle of the washed gas by displacement of air. (Hydrogen sulphide is a little heavier than air.) Will it burn? Will it support combustion? Has it any color or odor? Test its solubility in the water in the following way: Pass a few bubbles of the gas into water in a test-tube. Is the gas entirely absorbed? Is enough absorbed to impart its odor to the water? Complete the reaction showing the formation of hydrogen sulphide:

$$FeS + H_2SO_4 = \ldots$$

(b) *Formation of Sulphides by Direct Combination.*

Pass hydrogen sulphide into a solution of copper sulphate ($CuSO_4$). Write the reaction. Filter off the precipitated copper sulphide. See note on filtration, Exercise 9. Wash the precipitate by filling the cone formed by the filter-paper full of distilled water. The water, as it soaks through, dissolves out foreign matter held by the precipitate, thus "washing" it. Wash thoroughly. Is the precipitate soluble in dilute sulphuric acid? If it were, could it be formed by the reaction that you have written?

Using a solution of lead nitrate, precipitate lead sulphide. Filter and wash the precipitate. Test its solubility in dilute nitric acid.

Pass hydrogen sulphide into a solution of ferrous (iron) sulphate. Do you notice any precipitate or, at most, anything more than a slight precipitate? Now add ammonium hydrate and continue to pass in the gas. Filter and wash the precipitate. Reaction:

$$FeSO_4 + H_2S + 2NH_4OH = FeS + (NH_4)_2SO_4 + 2H_2O.$$

Test the solubility of the washed precipitate in dilute sulphuric acid. What odor is given off? Reaction? Why is ferrous sulphide not precipitated without the ammonium hydrate according to the reaction: $FeSO_4 + H_2S = FeS + H_2SO_4$?

Make a little ferrous sulphide as in Exercise 4, paragraph (*f*). Test it with dilute sulphuric acid. Is it the same substance that you have just studied? Why do you think so?

(*c*) *Test for Soluble Sulphides.*

Put a little sodium sulphide on a clean silver coin. Moisten with a drop of water and crush the sulphide. Allow to stand for some time and then wash. Note the black spot (silver sulphide) on the coin. Since sulphur or any of its compounds may be converted into sodium sulphide by fusing with sodium carbonate on charcoal, the above test is used to detect sulphur in any of its forms. Fuse a little sulphur mixed with an equal quantity of sodium carbonate on charcoal. Cut out the fused mass and test on silver as above.

(d) *Test for Hydrogen Sulphide.*

Moisten a piece of filter-paper with a drop of lead acetate solution. Hydrogen sulphide turns the paper black. Explain.

Has hydrogen sulphide acid, neutral, or basic properties?

References: Clk., pp. 136, 137; Cly., pp. 104, 105; R., pp. 231-234; S., pp. 160-163; S. and L., pp. 106-109; W., pp. 243-246.

EXERCISE 27.

SULPHUR COMPOUNDS CONTAINING OXYGEN.

(a) *Sulphur Dioxide.*

Burn a little sulphur (deflagrating-spoon) in a wide-mouthed bottle. Cautiously note the odor. Write the reaction. Add a little water to the bottle and shake well. Test the liquid with litmus paper.

Set up apparatus as shown in Fig. 15. Put 10 grms. of copper-foil or copper turnings in the flask and add about 25 cc. concentrated sulphuric acid through the thistle-tube. Heat gradually, increasing the temperature, till a gas comes off, then moderate the heating. Catch a bottle of the gas by displacement of air. Is it the same compound as you prepared by burning sulphur? Test the bleaching action of sulphur dioxide on a piece of red calico on a red flower, if you can procure one. Compare with the action of chlorine. Which bleaches the more vigorously?

Dissolve 1 grm. of sodium hydrate in 10 cc. of distilled water, and pass sulphur dioxide into the solution till the reaction is neutral to litmus. You now have a solution of sodium sulphite, Na_2SO_3. Write the reaction. To a small portion of the solution add HCl. What gas is given off? Write the reaction.

To another portion add barium chloride solution. Barium sulphite ($BaSO_3$) is formed. Reaction? Is barium sulphite soluble in dilute hydrochloric acid? (A little insoluble barium sulphate may be present.)

To another portion of the sodium sulphite solution add nitric acid and boil. Now add barium chloride solution and treat as above. Is the precipitate soluble in hydrochloric acid?

Nitric acid is a good oxidizing agent. You now have barium sulphate, which is insoluble in acids.

(b) *Sulphuric Acid.*

Describe the process of making sulphuric acid. Give some of the uses of this acid.

To 1 cc. water in a test-tube add 1 cc. concentrated sulphuric acid and note any change in temperature. Write with this acid on a piece of paper by means of a splinter. Warm for a few minutes and note result.

Put $\frac{1}{2}$ grm. sugar in a test-tube, add concentrated sulphuric acid and warm. What do you notice? What do you conclude from these experiments concerning the affinity of sulphuric acid for water?

To a drop of dilute sulphuric acid add a little barium chloride solution. Reaction? Try to dissolve the precipitate in water. In dilute acids. Results? Try the same experiments, using calcium chloride instead of barium chloride.

References: Clk., pp. 138–149; Cly., pp. 106–111; R., pp. 235–241; S., pp. 164–176; S. and L., pp. 110–120; W., pp. 124–132, 247, 248.

EXERCISE 28.

PHOSPHORUS.

For precautions in using phosphorus see Exercise 15.

(a) *Yellow Phosphorus.*

Put some water in an evaporating-dish and ask your instructor to place in it three pieces of phosphorus about the size of grains of wheat. Remove one with the pincers, dry by bringing in contact with filter-paper, and place in a clean dry porcelain crucible. Note that white fumes are given off. The phosphorus glows in the dark. What does the word *phosphorus* signify? Let the crucible containing the phosphorus float on water in a beaker. Place the beaker on the wire gauze and heat gently. Does the phosphorus melt? Does it burn? Note the temperature of the water with your hand. The white fumes are phosphorus pentoxide (P_2O_5). Write the reaction representing the burning of phosphorus.

Perform the following experiment exactly as directed. In a dry test-tube held with a clamp, place 1 cc. carbon disulphide. (Cork and remove from your working place any carbon disulphide that you may have aside from the amount just given.) Drop into the tube one piece of phosphorus. Does it dissolve in the carbon disulphide? Cover a beaker with a round filter a little larger in diameter than the beaker. Now, holding the tube with the clamp, pour the solution on the filter-paper. Stand a few feet distant and note what takes place after the carbon disulphide has evaporated.

Put a few crystals of iodine in a dry test-tube and drop on them a piece of dry phosphorus. The phosphorus unites with the iodine forming PI_2. Write the reaction.

What do you conclude from these experiments about the chemical activity of yellow phosphorus? Explain the use of phosphorus in matches.

(b) *Red Phosphorus.*

(This kind of phosphorus may be safely handled without special precautions.) Is red phosphorus soluble in carbon disulphide? Does it ignite at a low temperature? Try its action on iodine.

Put $\frac{1}{2}$ grm. of red phosphorus on the iron base of your ring-stand. Ignite under the hood. Do you think that the product of combustion is the same as when yellow phosphorus burns? Why? Compare the allotropic forms of phosphorus.

(c) *Phosphine.*

Heat a little calcium hypophosphite in a test-tube. Note the inflammability of the impure phosphine (PH_3) formed.

Complete the reaction: $2PH_3 + 4O_2 = \ldots$

(d) *Phosphoric Acid.*

Note.—The acids containing phosphorus are very numerous. We will here study only the orthophosphoric acid. Commercial phosphoric acid usually consists of metaphosphoric acid. This as well as pyrophosphoric acid, which is sometimes present, is converted into the ortho acid by boiling for some time with excess of water.

Neutralize a few cc. of a solution of orthophosphoric acid with ammonium hydrate. Be sure that the solution is exactly neutral to litmus. Add silver nitrate solution, describe the precipitate and write the reaction.

To 1 cc. of phosphoric acid solution add a few drops of a solution of ammonium molybdate. In very dilute solutions the precipitate appears after some time has elapsed. Describe the precipitate.

Repeat the experiment just given, using disodium phosphate instead of phosphoric acid. Result?

Test for phosphates in bone ash as follows: Digest 1 grm. of bone-ash with dilute sulphuric acid by gently heating (not boiling) for about 10 minutes. Filter and add ammonium molybdate solution to the clear filtrate. Do you obtain a precipitate? If so, describe it.

Try the same experiment, using coal-ash instead of bone-ash. Have you any evidence that coal-ash contains phosphates?

What use is made of the natural phosphates?

References: Clk., pp. 151–158; Cly., pp. 130–133; R., pp. 241–248; S., pp. 193–196; S. and L., pp. 139–147; W., pp. 249–258.

EXERCISE 29

ARSENIC.

Note.—Remember the poisonous properties of arsenic compounds.

Examine a piece of "metallic" arsenic about the size of a grain of wheat. Heat it on charcoal before the blowpipe. Note that a white deposit of arsenic trioxide appears on the charcoal. Write the reaction.

Make a small ignition-tube according to directions given in Exercise 6. Place in it a small amount (equal in volume to a coarsely broken half pea) of arsenic trioxide, "white arsenic." Dry a piece of charcoal by heating over the flame of a burner. Cool and cut it into the form of a cylinder 2 cm. ($\frac{3}{4}$ inch) long, having just the diameter of the bore of the tube. Push this charcoal plug into the ignition-tube to within about 13 mm. ($\frac{1}{2}$ inch) of the arsenic trioxide. Heat the charcoal as hot as possible, allowing the end of the tube containing the oxide to extend out of the flame. When the glass around the charcoal is nearly red-hot, hold the tube in a more nearly vertical position, so that the flame may strike both the charcoal and the white arsenic at the same time. What forms on the cool surface above the charcoal? What reaction has taken place between the vapor of arsenic trioxide and the carbon? When the tube is cool cut off the lower end and remove the carbon. Now hold the tube at an angle of 45° and heat again. What takes place? Explain fully.

(a) *Arsine* (AsH_3).

This compound is a deadly poison, hence it should not be prepared by any except experienced chemists. Study its properties in your text-book and describe it fully.

(b) *Comparison of Arsine with Stibine.*

Perform the following experiment under the hood.

Stibine or hydrogen antimonide (SbH_3) may be made as follows: Set up apparatus of the same general character as shown in Fig. 13. See that the small bottle is dry, and place in it granulated calcium chloride, in sufficient quantity to cover the bottom with a layer about an inch deep. The tube from the generator should reach almost to the bottom of the bottle. The calcium chloride serves to dry the gas. Attach at B a tube drawn out to a rather fine bore. Now generate hydrogen as directed in Exercise 8, and after all air is expelled light the gas. Hold in the flame a cool porcelain dish. Is a black spot produced? If so, arsenic or antimony may be present in the reagents. Now pour through the thistle-tube a few drops of a solution of any antimony compound (as tartar emetic). Does the hydrogen-flame change color? Hold the dish in the flame. Is a spot deposited? How could you distinguish it from an arsenic spot? (Consult text).

(e) Show some of the points of similarity between arsenic, antimony, and bismuth.

References: Clk., pp. 159-162, 247-252; Cly., pp. 134-137, 231; R., pp. 249-253; S., pp. 242-250, 255-258; S. and L., pp. 149-160; W., pp. 321-331, 336-339.

EXERCISE 30.

SILICON ; BORON ; REVIEW OF NON-METALS.

(a) *Silicon.*

How is silicon prepared? Give its chemical and physical properties. (Consult text.)

To a solution of sodium silicate "water-glass" add dilute hydrochloric acid to distinctly acid reaction. Evaporate very cautiously to dryness in a small evaporating-dish on the wire gauze. The residue is silicon dioxide (SiO_2) mixed with common salt. Is it (SiO_2) soluble in water? in acids? Silicon dioxide (silica) is the anhydride of the silicic acids. One of these has the symbol H_2SiO_3. Compare its structure with that of carbonic acid.

What is quartz? Quartz sand? Glass? How is glass made?

(b) *Boron.*

How is boron prepared? Give its physical and chemical properties. (Consult text.)

Dissolve 10 grms. of borax (sodium biborate, $Na_2B_4O_7$) in 25 cc. boiling water. To the hot solution add 6 cc. concentrated hydrochloric acid. On cooling crystals of boric acid (H_3BO_3) separate out. Write the reaction.

Filter off the boric acid from the mother liquor, and dissolve a few crystals in 2 cc. of alcohol. Set fire to the solution in a small evaporating-dish. Note the color of the flame.

(c) *Review.*

Explain the difference between metals and non-metals. Is it possible to make a sharp distinction in all cases?

Make a list of all the non-metallic elements that you have studied, giving the atomic weights. Also make a list of the compounds that you have studied, giving formulas and molecular weights. (This list is not to include complicated compounds, as sugar, starch, starch iodide, etc.)

References: Clk., pp. 163–171; Cly., pp. 137, 138, 143–147; R., pp. 255–260; S., pp. 184–192; S. and L., pp. 161–165; W., pp. 259–275.

THE MORE IMPORTANT METALS.

EXERCISE 31.

THE ALKALI METALS.

Make a list of the metals of this group, arranged in the order of their atomic weights.

(a) Sodium.

Put a few cc. of kerosene in a perfectly dry dish, and ask your instructor to place in it a piece of sodium as large as a small pea. Examine the metal. (Handle alkali metals with pincers.) Describe its physical properties. Cut off a little, free from kerosene by pressing between filter-paper, and expose to the action of the air. After twenty minutes examine the piece What do you notice? Why is sodium kept under kerosene?

Pour into a small evaporating-dish about 10 cc. of distilled water. Now drop in about half the sodium that you have and cover immediately with a glass plate. (This is to protect your eyes in case there should be a slight explosion at the end of the reaction.) What do you observe? Heat the solution thus obtained nearly to boiling and add the rest of the sodium. What takes place? To what is the flame due? To what is the color of the flame due?

Test the solution, obtained by dissolving the two portions of sodium, with red litmus paper. Now add hydrochloric acid till the solution is exactly neutral. Evaporate to dryness and taste a little of the residue.

What is it? Write the reactions, showing how you have made salt from sodium.

(b) *Potassium.*

Ask your instructor for a piece of potassium about the size of a grain of barley. Keep under kerosene till you are ready to use it. Examine its physical properties. Now drop it on cold water, covering the dish as in the case of sodium. What do you notice? Which do you regard as the most active chemically, sodium or potassium? Which has the greater atomic weight? Which would you expect to find the more active chemically, lithium or cæsium? Why?

(c) *Solubility of the Salts of the Alkali Metals.*

The following may be taken as representative compounds: sodium sulphate, sodium carbonate, sodium phosphate, potassium nitrate, potassium chloride, potassium bichromate. Take six perfectly clean and dry test-tubes and place in each as much of one of the above-mentioned salts as will stand on the point of a knife-blade. Add to each tube a little water, and warm. Do you find them all soluble? What do you conclude in regard to the solubility of most of the alkali compounds?

(d) *Flame Reactions.*

Procure from your instructor a piece of platinum wire about $2\frac{1}{2}$ inches long. Clean it before each test, using a clean cloth moistened with distilled water.

Place one drop of lithium chloride solution in a clean dish. Moisten the wire in this solution and hold in the colorless flame of the burner. Note color of flame. Try the same experiment with potassium chloride; with sodium chloride; with any sodium compound.

Name some of the salts of sodium and of potassium that are of commercial importance.

References: Clk., pp. 183–192; Cly., pp. 216–218; R., pp. 283–302; S., pp. 320–336; S. and L., pp. 288–315; W., pp. 280–287.

EXERCISE 32.

COMPOUNDS OF THE ALKALINE EARTHS (CALCIUM, STRONTIUM, AND BARIUM).

(a) *Calcium Compounds.*

To 2 grms. of precipitated chalk (calcium carbonate, $CaCO_3$) add dilute hydrochloric acid, drop by drop, with constant stirring, till solution is just effected. Now add a little more chalk to neutralize any excess of acid. Heat and stir well in order that all of the acid may come in contact with chalk. Add a little water to replace what has evaporated, and filter from the undissolved portion. Complete the reaction:

$$CaCO_3 + 2HCl = CaCl_2 + \ldots$$

Test the flame reaction of the solution of calcium chloride just prepared. To a few drops of the calcium chloride solution add an equal volume of sodium carbonate solution. The precipitate is calcium carbonate. Write the reaction. Try the action of ammonium carbonate. Reaction? To the rest of the solution of calcium chloride add sodium hydrate solution. Calcium hydrate is precipitated. Write the reaction.

Fill a large test-tube half full of a clear solution of calcium hydrate (lime-water). Generate carbon dioxide by the action of hydrochloric acid on marble, using the apparatus shown in Fig. 13. The gas is to be washed with water in the small bottle in order to free it from traces of hydrochloric acid. The gas is now

COMPOUNDS OF THE ALKALINE EARTHS. 89

passed into the lime-water till the precipitate which forms at first is for the most part redissolved. Filter the solution into a clean beaker and boil for a few minutes. What do you notice? Is this deposit calcium carbonate? Why do you think so? We will now study the reactions involved in the entire process. You have already learned that calcium carbonate is formed by the action of carbon dioxide on lime-water. The next step is the solution of calcium carbonate according to the reaction:

$$CaCO_3 + CO_2 + H_2O = CaH_2(CO_3)_2.$$

This compound (acid calcium carbonate) remains in solution till heated, when it breaks up according to a reaction that is just the reverse of the one last given. Write this reaction. Explain how water containing carbon dioxide dissolves calcium carbonate from the soil and rocks, becomes hard, and leaves a deposit of "lime" in steam-boilers and tea-kettles.

(b) Strontium Compounds.

Test the flame reaction of strontium chloride.

To a few drops of strontium chloride solution add the same volume of sodium carbonate solution. Complete the reaction: $SrCl_2 + Na_2CO_3 = \ldots$

To another portion of the solution add a little dilute sulphuric acid. Complete the reaction:

$$SrCl_2 + \ldots = SrSO_4 + \ldots$$

To a third portion add sodium hydrate solution. Strontium hydrate $Sr(OH)_2$ is precipitated. Write the reaction.

(c) *Barium Compounds.*

Perform the same experiments with barium chloride that you did with strontium chloride. Describe results and write reactions. (The reactions are analogous.) Ba is substituted for Sr.)

The formation of a precipitate indicates the formation of an insoluble or a difficultly soluble compound. What, then, do you conclude as to the solubility of most of the compounds of the alkaline earths?

References: Clk., pp. 205–212; Cly., pp. 220, 221; R., pp. 303–318; S., pp. 308–316; S. and L., pp. 315–325; W., pp. 288–293, 143–147.

EXERCISE 33.

MAGNESIUM, ZINC, CADMIUM, AND MERCURY.

In each of the groups of metals that we have studied we have found a striking similarity of properties. The grouping of the rest of the metals is more or less arbitrary. Thus the metals considered in this exercise do not resemble each other in so marked a degree as do the alkaline earths.

(a) Magnesium.

Describe the metal. Treat a little of the metal (a piece of magnesium ribbon 6 mm. long) with dilute hydrochloric acid. Try the same experiment with dilute sulphuric acid. Magnesium chloride ($MgCl_2$) and magnesium sulphate ($MgSO_4$) are formed respectively. Write the reactions.

Hold with the pincers a piece of magnesium ribbon about 25 mm. (1 inch) long in the flame of the burner till the magnesium is ignited. Magnesium oxide is formed. Write the reaction. Prepare a bottle of carbon dioxide and thrust into it a piece of burning magnesium ribbon. Does the burning continue? Are the products of combustion the same as when magnesium burns in the air? This experiment shows the affinity of magnesium for oxygen.

In a piece of charcoal make a hole large enough to hold a half pea. Fill with magnesium oxide slightly moistened with water. Now heat strongly before the blowpipe, cool, moisten with one drop of cobaltous nitrate, and heat again. Do you obtain a red-brown color? This is a test for magnesium oxide.

(b) *Zinc.*

Put a little zinc in a test-tube and treat with dilute hydrochloric acid. Repeat, using dilute sulphuric acid. Reactions? Heat a little zinc on charcoal in the oxidizing-flame of the blowpipe. Note the white coating of zinc oxide (ZnO) on the charcoal. Reaction?

Treat a little zinc oxide just as you did magnesium oxide. Note the color when hot, when cold, after treatment with cobaltous nitrate and heating a second time.

How is zinc obtained from its ores?

(c) *Cadmium.*

Procure from your instructor three pieces of cadmium. (The entire amount need not exceed $\frac{1}{2}$ grm. in weight.) Heat one piece in an ignition-tube. Does it melt at a comparatively low temperature? Allow to cool and then put on charcoal and heat in the oxidizing-flame. What do you notice? What is the color of cadmium oxide (CdO)?

Test a piece of cadmium with dilute hydrochloric acid. Cadmium chloride ($CdCl_2$) is formed. Reaction?

Test another piece with dilute sulphuric acid. Reaction? To a few drops of a solution of cadmium chloride add ammonium sulphide. What is the color of the cadmium sulphide (CdS).

(d) *Mercury.*

Give the physical properties of mercury. At what temperature does it melt? Try the action of dilute hydrochloric acid on a drop of mercury. Does the metal dissolve?

To a little mercurous nitrate ($HgNO_3$) solution add a little dilute hydrochloric acid: mercurous chloride ($HgCl$) is precipitated. Write the reaction. Repeat the same experiment, using mercuric nitrate ($Hg(NO_3)_2$) solution. Is a precipitate formed? Which, then, is soluble in water, mercurous chloride (calomel) or mercuric chloride (corrosive sublimate)?

Place in a few cc. of any mercury salt solution a piece of bright copper; withdraw after a few minutes and rub with a cloth. Explain what you observe.

Make a list of the elements studied in this exercise in the order of their atomic weights. Give their atomic weights. Compare the metals in respect to the following properties: specific gravity, melting-point, ease with which they are attacked by acids.

Show in what relation these properties stand to the atomic weights.

References: Clk., pp. 223–231; Cly., pp. 223, 224, 236–237; R., pp. 319–324, 330, 331; S., pp. 316, 317, 298, 300, 262, 263, 233, 237; S. and L., pp. 325–337, 347–350; W., pp. 301–305, 339, 340, 356–358.

EXERCISE 34.

COPPER, SILVER, AND GOLD.

(a) Copper.

To a solution of copper sulphate add a little cold sodium hydrate solution. Copper hydrate ($Cu(OH)_2$) is formed. Write the reaction. Add to a solution of copper sulphate a drop or two of dilute ammonia. What do you notice? Now add excess of ammonia. What color is produced? To 2 or 3 cc. of water add a drop of copper sulphate solution, then a drop of potassium ferrocyanide solution. Note the color of the copper ferrocyanide. This is a very delicate test for copper. To a few drops of copper sulphate solution add 1 cc. of a solution of tartaric acid, then sodium hydrate solution till the solution is strongly basic. Now add a little grape-sugar and boil. Cuprous oxide (Cu_2O) is precipitated. Describe it. To a little copper sulphate solution add a drop of dilute hydrochloric acid, then a piece of bright iron (wire nail). Examine after a few minutes. How do you account for the appearance of the nail?

Heat a little bright copper in the flame of the burner. Does it oxidize? Test the solubility of copper in dilute nitric acid. Write the reaction. (See Exercise 17.)

What use is made of copper in the arts? What are some of the most important alloys of copper? How is copper found in nature?

(b) Silver.

To a little silver nitrate solution add an equal volume of dilute hydrochloric acid. Reaction? Pour off the liquid and dissolve the precipitate in ammonia. Now add nitric acid. What appears? Collect the precipitate on a very small filter, wash with distilled water and allow to drain well. Now add to the precipitate its own volume of dry sodium carbonate, and roll the filter and contents into a ball (size of a pea). Heat this on charcoal before the blowpipe till the paper is consumed. If the silver appears in scattered deposits, scrape them together and fuse into a single bead. Does silver oxidize when heated in the air? Try the solubility of the silver bead in nitric acid. Why is silver classed among the noble metals? What is "coin silver"? "sterling silver"?

(c) Gold.

See text-book for the properties of gold.

Does gold oxidize when heated? Does it dissolve in any simple acid? Compare it with silver and with copper in these respects. What is the composition of the alloy of which gold coins are made? What is meant by the term 18-carat gold?

References: Clk., pp. 275–279, 196–203, 280–281; Cly., pp. 233–236, 237–245; R., pp. 325–329, 332–336, 378–381; S., pp. 258–261, 228–233, 266–267; S. and L., pp. 341–346, 350–356, 383–385; W., pp. 340–343, 349–356, 369–361.

EXERCISE 35.

ALUMINIUM (OR ALUMINUM).

Procure two small pieces of the metal from your instructor. Test the solubility (using one piece) with dilute hydrochloric acid. Aluminium chloride ($AlCl_3$) remains in solution. Write the reaction. Test the solubility of aluminium in dilute sodium hydrate solution. Sodium aluminate ($NaAlO_2$) remains in solution. Write the reaction. Describe the physical properties of aluminium.

To a solution of aluminium sulphate ($Al_2(SO_4)_3$) add ammonium hydrate. Aluminium hydrate ($Al(OH)_3$) is precipitated. Write the reaction. Filter and heat the precipitate on charcoal before the blowpipe, moisten with a little cobaltous nitrate and heat again. Note the color. This is a test for aluminium.

Alum is a good example of a "double salt."

Dissolve with the aid of heat 10 grms. of aluminium sulphate in as little water as will effect solution. Now dissolve 2.6 grms. of potassium sulphate in the same manner and mix the hot saturated solutions of the two salts. Crystals of alum ($AlK(SO_4)_2.12H_2O$) will form on cooling. Filter off the crystals from the mother-liquor and show to the instructor. How can you show that they contain aluminium? (See reactions above.) How can you prove that they contain potassium? (See flame-test, Exercise 30.)

How is aluminium reduced from its ores? Give uses of the metal; of its alloys.

References: Clk., pp. 232-236; Cly., pp. 221-223 R., pp. 339-348; S., pp. 286-290; S. and L., pp. 357-363; W., pp. 313-317.

EXERCISE 36.

TIN AND LEAD.

(a) Tin.

Give the physical properties of the metal. Put about ½ grm. of granulated tin into a test-tube and add hydrochloric acid. The tin dissolves in the acid, forming a solution of stannous chloride ($SnCl_2$). After the greater part of the tin has dissolved, pour off the solution from the undissolved metal. To this solution add a few drops of mercuric chloride solution. Now heat. The precipitate varies from white to grayish black, depending on the temperature and the proportion of each reagent added. The reactions may be expressed as follows:

$2HgCl_2 + SnCl_2 = 2HgCl$ (white ppt.) $+ SnCl_4$ (sol.).
$2HgCl + SnCl_2 = 2Hg$ (dark gray) $+ SnCl_4$ (solution).

This is a delicate test for both tin and mercury.

To a little tin add moderately concentrated nitric acid. What do you notice? Metastannic acid ($H_2SnO_3)_x$ is formed. Is it soluble in water? Filter, wash and drain well. Now heat a little in a porcelain crucible to a high temperature. Stannic oxide is formed. Write the reaction.

What use is made of tin in the arts? What is bronze?

(b) Lead.

Dissolve ½ grm. of lead in nitric acid in the following manner: Place the lead in an evaporating-dish containing a few cc. of water. Heat to boiling and add concentrated nitric acid till vigorous action takes place. Now set under the hood and allow to stand till action nearly ceases. If undissolved lead is present, add a little more acid and warm. If this does not cause action to begin again, add a little more water. In order to dissolve lead readily there must be acid enough to unite with the lead, and water enough to hold in solution the lead nitrate formed. When solution is effected, evaporate nearly to dryness and dissolve the residue in 500 cc. of water. Suspend in this solution a strip of zinc and allow it to remain for an hour. What do you notice? Complete the reaction:

$$Zn + Pb(NO_3)_2 = Pb + \ldots$$

To a solution of lead nitrate ($Pb(NO_3)_2$) add dilute sulphuric acid. Write the reaction. To another portion of the nitrate solution add a solution of potassium chromate (K_2CrO_4). Lead chromate is formed. Write the reaction.

Warm a little red lead (Pb_3O_4) with dilute nitric acid in excess. Brown lead peroxide (PbO_2) is left undissolved.

How is lead reduced from its ores? What use is made of lead in the arts? What are some of the more important alloys of lead?

References: Clk., pp. 239–245; Cly., pp. 230, 232, 233; R., pp. 349–357; S., pp. 251–254, 224–228; S. and L., pp. 381–383, 337–341; W., pp. 331–335, 334–349.

EXERCISE 37.
CHROMIUM.
(a) Chromates.

In these compounds chromium has an acid character. Chromic acid (hydrogen chromate (H_2CrO_4)) is known only in solution. The salts of this acid are similar in structure and crystalline form to the salts of sulphuric acid.

Dissolve a few crystals of potassium chromate in water, and to separate portions add in solution: silver nitrate, barium chloride, lead nitrate. Reaction for silver nitrate:

$$K_2CrO_4 + 2AgNO_3 = 2KNO_3 + Ag_2CrO_4.$$

Now write the other two reactions, remembering that one atom of lead or barium can replace two atoms of potassium.

To about 1 cc. of potassium chromate solution add a few drops of any dilute acid on your desk. Note change of color. Potassium bichromate is formed. Complete the following:

$$2K_2CrO_4 + 2HCl = K_2Cr_2O_7 + \ldots$$

To this solution add a solution of potassium hydrate till the color just changes to yellow. Potassium chromate (K_2CrO_4) is again formed. Write the reaction.

(b) Chromic Compounds.

When chromates are reduced, i.e. lose oxygen, the acid character of chromium disappears. Chromium in chromic compounds has a basic character.

To about 1 cc. of a solution of potassium bichromate add a little concentrated hydrochloric acid and a few drops of alcohol. Warm and note change of color. Repeat, using a little oxalic acid instead of alcohol.

To a solution of chrome alum add ammonia. The reaction is as follows:

$$CrK(SO_4)_2 + 3NH_4OH$$
$$= KNH_4SO_4 + (NH_4)_2SO_4 + Cr(OH)_3.$$

Mix a little ($\frac{1}{2}$ grm.) dry sodium carbonate with an equal quantity of potassium nitrate and half as much chrome alum. Heat strongly in a porcelain crucible for some time, cool and dissolve in water. If the oxidation is complete the solution will appear yellow. To the solution add dilute nitric acid till it just begins to turn orange-red. Add now a few drops of silver nitrate solution; a red-brown precipitate shows the presence of a chromate.

Fuse a piece of platinum wire into a piece of glass tube; the shape is shown in Fig. 22.

FIG. 22.

Heat the wire loop in the flame, and bring it into contact with a small crystal of borax. Fuse this to a globule, continuing the process till you have a clear bead of borax glass just large enough to be held in the loop.

Fuse into this bead a minute quantity of chrome alum, just enough to give a decided color, not enough to make it opaque. What do you notice? This is a delicate test for chromium.

References: Clk., pp. 252-256; R., pp. 372-375; S., pp. 282-285; S. and L., pp. 378, 379; W., pp. 317-320.

EXERCISE 38.

MANGANESE.

(a) Oxidizing Action of Permanganates.

Dissolve a crystal of potassium permanganate ($KMnO_4$) in 10 cc. of water. Place in a beaker and heat nearly to boiling. To the hot solution add, a drop at a time, a solution of oxalic acid containing a little sulphuric acid. Is the color discharged? Repeat, using a solution of ferrous sulphate containing sulphuric acid in place of the oxalic acid solution.

To a few crystals of potassium permanganate in an evaporating-dish add a few drops of concentrated sulphuric acid. Put 1 cc. of strong alcohol in a dry beaker and pour it upon the mixture just made. What do you notice?

To a few crystals of potassium permanganate in a test-tube add concentrated hydrochloric acid. Is the acid oxidized? If so, what gas is liberated? How do you recognize it?

(b) Manganous Compounds.

To a solution of manganous sulphate add sodium hydrate. Manganous hydrate is formed. Complete the reaction: $MnSO_4 + 2NaOH = \ldots$

Allow the precipitate to stand exposed to the air till it changes in color. Manganic hydrate ($Mn(OH)_3$) is formed.

To a solution of manganous sulphate add a solution of ammonium sulphide. Manganous sulphide (MnS) is precipitated. Describe precipitate and write reaction.

Test any manganese compound in the borax bead. (See Exercise 37.) What color is imparted to the borax glass?

What are some of the uses of manganese compounds?

References: Clk., pp. 262-264; R., pp. 369-371; S., pp. 295-298; S. and L., pp. 363-365; W., pp. 298-301.

EXERCISE 39.

IRON, COBALT, NICKEL, AND PLATINUM.

(a) *Iron in Ferrous Compounds.*

Put about 3 grms. of iron turnings into a test-tube and add, a little at a time, about 10 cc. dilute hydrochloric acid. Ferrous chloride ($FeCl_2$) is formed. Reaction. Allow to stand till action nearly ceases, then pour off the solution, about 1 cc. at a time, and test its action with solutions of the following reagents: (1) sodium hydrate; (2) potassium ferrocyanide; (3) potassium ferricyanide. Describe results. Write reaction for (1). Allow this precipitate to stand exposed to the air. Ferric hydrate ($Fe(OH)_3$) is formed.

(b) *Iron in Ferric Compounds.*

Try the action of ferric chloride ($FeCl_3$) solution on reagents (1), (2), and (3). Describe results and write reaction for (1). To about 1 cc. of a solution of ferrous chloride add a few drops of concentrated nitric acid and boil. The solution should now be light yellow in color. If it is green or dark, repeat the process. Avoid an excess of acid. Now prove by the action of reagents (1), (2), and (3) that ferrous iron has been oxidized to ferric iron.

What are the most important ores of iron? What is cast iron? wrought iron? steel?

(c) *Cobalt.*

To 1 cc. of a solution of cobaltous chloride add sodium hydrate solution and boil. Cobaltous hydrate is precipitated. Write the reaction. (See analogous reation of $FeCl_2 + 2NaOH$.) What color is imparted to the borax bead by cobalt compounds?

(d) *Nickel.*

Treat a solution of nickel sulphate ($NiSO_4$) with sodium hydrate solution. What is formed? Reaction. What color is imparted to the borax bead by nickel compounds?

What use is made of nickel in the arts? What are some of the most important alloys of nickel? Of what does "nickel" used for coinage consist?

(e) *Platinum.*

What properties of platinum have you observed? Why is it valuable for use in the laboratory? How is it found in nature?

References: Clk., pp. 264–271, 273, 274, 282–284; Cly., pp. 224–230, 244; R., pp., 358–368, 376–378; S., pp. 275–282, 290–295, 268, 269; S. and L., pp. 366, 378, 385–388; W., pp. 306–313, 294–298, 361.

SOME FAMILIAR HYDROCARBONS AND THEIR DERIVATIVES.

SO-CALLED ORGANIC COMPOUNDS.

EXERCISE 40.

SOME HYDROCARBONS.

(a) The Methane Series.

This series has the general formula C_nH_{2n+2}. It is rather difficult to prepare the lower members. Consult text and reference-books for the preparation and properties of the first member, methane (CH_4). Examples of the higher members are benzine, gasoline, kerosene, and paraffine.

The greater part of gasoline consists of heptane (C_7H_{16}) and octane (C_8H_{18}). Place four drops of gasoline in an evaporating-dish. (Keep bottle of gasoline away from flames.) Place the dish under the hood and apply a lighted splinter. Note the character of the flame.

Now repeat the same experiment, using kerosene instead of gasoline. (Kerosene consists principally of hydrocarbons, $C_{12}H_{26}$ to $C_{16}H_{34}$ inclusive.) Which can you light the most readily, gasoline or kerosene? Which burns with the higher flame? Which flame is the more sooty? What relation does the number of carbon atoms bear to the volatility of the hydrocarbon?

When any hydrocarbon is completely burned, water and carbon dioxide are the only products formed. Write the reaction representing the combustion of hexane (C_6H_{14}). What is the composition of American petroleum? (See reference-books or cyclopedia.)

(b) *Acetylene* (C_2H_2).

This hydrocarbon belongs to the series C_nH_{2n-2}.

Fill a test-tube one fourth full of water and drop into it two or three pieces of calcium carbide (CaC_2) as large as peas. Light the escaping gas. Note the appearance of the flame. Write the reaction representing the combustion of acetylene.

After action in the tube has ceased test the contents of the tube with red litmus paper. Have you reason to think that calcium hydrate ($Ca(OH)_2$) is present? If so, write the reaction representing the action of water on calcium carbide resulting in the liberation of acetylene.

References: Clk., pp. 79–83, 296, 297; Cly., pp. 183–186; R., pp. 384–389; S., pp. 132–135; S. and L., pp. 196–206, 238, 239; W., pp. 363–370.

EXERCISE 41.

SOME HALOGEN DERIVATIVES OF THE HYDROCARBONS.

(a) *Chloroform* ($CHCl_3$).

Chloroform may be made by treating methane with chlorine. It is regarded as methane in which three chlorine atoms have been substituted for three hydrogen atoms.

Pour about 3 cc. of chloroform into a test-tube provided with a cork stopper, and use for the following tests: Put a few drops into a dry test-tube and warm. Note the odor. Will the vapor burn? To a few cc. of water in a test-tube add a few drops of potassium iodide solution and a minute crystal of iodine. When solution is effected add a few drops of chloroform to the colored solution, shake vigorously, and allow to settle. What do you notice? Is chloroform a good solvent for iodine? for gums and oils? Try its action on a little paraffine. What is the principal use of chloroform?

(b) *Iodoform* (CHI_3).

Place in a test-tube 1 cc. of ordinary (ethyl) alcohol and add 2 grms. of iodine. Now dissolve in a small beaker or test-tube about 3 grms. of potassium carbonate (or sodium carbonate) in a small quantity of hot water. Add this to the iodine in the alcohol, drop by drop, till the brown color disappears. Boil the mixture till any pieces of iodine that may be in the bottom of the tube are dissolved. If the solution becomes brown, add a drop or two more of the carbonate solution. Now heat to boiling, set aside, and allow to cool. Yellow crystals of iodoform should separate out. For what is iodoform used?

References: Clk., pp. 302; R., pp. 387, 388; S. and L., pp. 197, 198; W., pp. 366.

EXERCISE 42.

ALCOHOL.

An alcohol may be defined as a hydrocarbon in which hydrogen (H) has been replaced by hydroxyl (OH).

(a) Methyl Alcohol (CH_3OH).

Procure 1 cc. of methyl alcohol (or wood spirit) and note odor. Draw some of the vapor (not the liquid) into the mouth and note the taste. Do you think that this alcohol would admit of use in beverages? What is "methylated spirit"? Will methyl alcohol burn? If so, write the reaction.

(b) Ethyl Alcohol (C_2H_5OH).

Put 20 grms. of commercial glucose or 30 grms. of "corn-syrup" into a 250-cc. flask, add about 150 cc. of water and one fourth of a cake of "compressed" yeast. Fermentation should begin promptly, according to the reaction:

$$\underset{\text{Grape-sugar}}{C_6H_{12}O_6} = \underset{\text{Alcohol}}{2C_2H_5OH} + \underset{\text{Carbon dioxide}}{2CO_2}.$$

To prove that carbon dioxide is given off conduct the escaping gas into lime-water as shown in Fig. 23. Cover the lime-water with a thin layer of kerosene to protect it from the action of carbon dioxide in the air. Be sure that the cork in the flask is *perfectly tight*, otherwise the gas will gradually diffuse into the air instead of passing into the lime-water. Set aside for *several* days. Does the lime-water become cloudy? If so, give the reaction. Now filter the contents of the flask. The filtrate consists of some unchanged glucose, alcohol, and traces of other products dissolved in a large excess of water. The alcohol may be obtained in purer condition and more concentrated form by distilling.

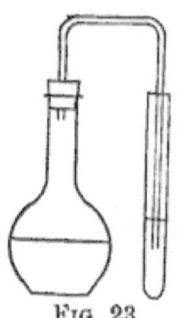

Fig. 23.

Set up apparatus as shown in Fig. 18. Pour into the retort about one half as much of the filtrate as it will hold and then drop in a few pieces of granulated zinc to prevent "bumping."

Distil over, very cautiously, about one fourth of the liquid in the retort, keeping the test-tube used as a receiver well cooled. This distillate contains nearly all the alcohol of the liquid placed in the retort along with considerable water. Save the distillate; throw away the liquid remaining in the retort. Repeat the process till all the alcohol in the filtrate has been distilled off. If directions have been followed, you have now distilled off 40 to 50 cc. of dilute alcohol. Now wash the retort, and in it place this dilute alcohol. Distil off about 10 cc. You should now have the alcohol in a fairly concentrated form. Put 1 cc. of this alcohol into a test-tube, add a few crystals of iodine, and then potassium carbonate solution. See method of making iodoform in Exercise 40. The formation of iodoform shows the presence of alcohol.

Put the rest of the alcohol in a beaker and gradually heat to boiling. Apply a light to the vapor. If the alcohol does not contain too much water its vapor will burn. Write the reaction.

Note.—Conditions affecting Fermentation. The yeast must be fresh. "Compressed" yeast gives better results than dry yeast. Fermentation requires air at the start, hence do not make the test for carbon dioxide till the gas begins to come off freely. The proper temperature is 20°–30° C. (68°–86° F.). Sunlight is unfavorable. Disinfectants, as certain salts, carbolic acid, etc., kill the yeast-cells. Impure sugars or syrups contain albuminoids and phosphates, hence they ferment more readily than pure sugars.

References: Clk., pp. 298–300; Cly., pp. 186–189; R., pp. 389–391, 400, 401; S. and L., pp. 210–214; W., pp. 365, 366, 370, 371.

EXERCISE 43.

SOME FATTY ACIDS.

These acids are called "fatty acids" because some of the higher members, combined with glycerin, constitute a considerable portion of animal and vegetable oils. Hence the hydrocarbons from which they are derived are called "fatty" hydrocarbons.

(a) *Acetic Acid* ($H(C_2H_3O_2)$).

Set up apparatus as in Fig. 18. Place a few pieces of glass in the retort and add about 30 cc. of strong vinegar. Distil off the greater part of the liquid. Taste the distillate.

Try its action on blue litmus paper; on sodium carbonate solution. Complete the reaction:

$$Na_2CO_3 + 2H(C_2H_3O_2) = \ldots$$

To a little of the distillate add a few drops of ferric chloride solution. Note color of solution.

(b) *Soap.*

Soap consists principally of the sodium (or potassium) salts of the higher fatty acids. Thus we may assume that a pure white soap consists of sodium palmitate ($Na(C_{16}H_{31}O_2)$) and sodium stearate ($Na(C_{18}H_{35}O_2)$). Since this latter compound predominates we will use it in writing all reactions in this section.

Weigh out 5 grms. of any white soap ("Ivory" soap is good), and cut into fine shavings. Dissolve these in 100 cc. of distilled water, heating if necessary. If the solution is not perfectly clear, filter.

Make a saturated solution (5 cc.) of common salt,

and to this solution add an equal volume of the soap solution. What do you notice? Soap is not soluble in brine. Explain what use is made of this fact in soap manufacture.

To another portion of the soap solution add a few drops of magnesium sulphate. Result?

Complete the reaction $MgSO_4 + 2Na(C_{18}H_{35}O_2) =$

Is the precipitate soluble in water? Explain how magnesium sulphate makes water "hard."

Prepare a solution of acid calcium carbonate. (See Exercise 32.) Test a little of the solution with the soap solution. Is a precipitate formed?

The reaction is:

$$CaH_2(CO_3)_2 + 2NaC_{18}H_{35}O_2 = Ca(C_{18}H_{35}O_2)_2 + 2NaHCO_3.$$

Now boil the rest of the acid calcium carbonate solution for several minutes and filter. Test the filtrate with soap solution. Do you now get a precipitate? Explain. What is temporary hardness in water?

To the rest of the soap solution add hydrochloric acid to decidedly acid reaction. The precipitate consists of stearic acid (with other acids). Write the reaction. Is this acid soluble in water?

References: Clk., pp. 303–306, 319–321; Cly., p. 206; R., pp. 393–396; S. and L., pp. 215, 216, 224–227; W., pp. 372, 373.

EXERCISE 44.

SOME FAMILIAR CARBOHYDRATES.

(a) *Glucose or Grape-sugar* ($C_6H_{12}O_6$).

Prepare Fehling's test for glucose as follows: Dissolve about 1.8 grms. of crystallized copper sulphate in 25 cc. of water. Label this solution "I." Dissolve 8 grms. of Rochelle salt (sodium potassium tartrate) in 25 cc. of water. To this solution add 2.5 grms. of pure (solid) sodium hydrate and heat gently till solution is effected. Label this solution "II."

To 5 cc. of solution "II" add an equal quantity of solution "I." Place in a small beaker and heat to boiling. If the reagents are pure, no change will be noticed. Now add a little glucose and heat again. What do you notice? Cuprous oxide, containing more or less water, is precipitated.

(b) *Cane-sugar* ($C_{12}H_{22}O_{11}$).

Warm a few crystals of cane-sugar with a drop of concentrated sulphuric acid and note result. Test with Fehling's solution as in (a), using cane-sugar instead of glucose. If no precipitate is formed, test with a few drops of a solution of cane-sugar that has been boiled with a drop of concentrated hydrochloric acid. Describe results. Explanation: Under the influence of the acid we have

$$C_{12}H_{22}O_{11} + H_2O = C_6H_{12}O_6 + C_6H_{12}O_6.$$
Cane-sugar + Water = Grape-sugar + Fruit-sugar.

(c) *Starch* $(C_6H_{10}O_5)_x$.

How can you detect starch by means of iodine? (See Exercise 22.) Prove that a sample labelled "starch" really contains starch. Work out the details of the test for yourself. Describe results fully.

Does starch give a precipitate with Fehling's solution? Can you obtain a precipitate by first converting starch into glucose? Proceed as follows: Put a little starch into a test-tube and add a few drops of concentrated sulphuric acid. Allow to stand for a few minutes, then add a little water and boil.

Does a little of this reduce Fehling's solution?

(d) *Cellulose* $(C_6H_{10}O_5)_y$.

Filter-paper and absorbent cotton are examples of nearly pure cellulose. Is cellulose soluble in water or dilute acids? Can it be converted into glucose? Try as follows: Put a little filter-paper (a round filter 7 cm. in diam.) into a perfectly dry beaker and add 1 cc. of concentrated sulphuric acid. Using a glass rod, thoroughly stir the paper into a thick paste. Allow to stand for a few minutes, then add 2 cc. of water and boil. Nearly neutralize with solid sodium carbonate, adding a little water from time to time in quantity sufficient to dissolve the greater part of the sodium sulphate that is formed. Test this solution with Fehling's solution.

The following experiment is to be performed if ice or snow is procurable. Set a small beaker in a dish containing pieces of ice or snow. Add 10 cc. concentrated nitric acid and allow to cool. Now pour in 20 cc. concentrated sulphuric acid and stir with a glass rod. Allow the mixed acids to cool and then introduce a tuft of absorbent cotton equal in volume to that of the mixed acids. Allow to stand for 15 minutes, remove with a glass rod and wash thoroughly. Set aside and allow to dry. Compare it with absorbent cotton. Place an equal amount of absorbent cotton near it and apply a lighted splinter to each. What do you notice? For what is nitrocellulose used? What is gun-cotton?

Why are the compounds in this exercise called carbohydrates?

References: Clk., pp. 322–328; Cly., pp. 203, 204; R., pp. 402–408; S. and L., pp. 253–269.

EXERCISE 45.

A Few Aromatic or Benzene Derivatives.

(a) *Benzene or Benzol* (C_6H_6).

This compound is not to be confused with commercial "benzine."

Note the odor of benzene. Procure a few drops in an evaporating-dish and apply a lighted splinter. Note the appearance of the benzene flame.

(b) *Nitrobenzene* ($C_6H_5NO_2$).

Perform this experiment under the hood. See that no lighted burners are near. Avoid inhaling the fumes.

Put 5 cc. concentrated nitric acid into a small flask and add an equal volume of benzene. Now add, a drop at a time, 2 cc. concentrated sulphuric acid, stirring constantly. In case the mixture becomes hot so as to seem to boil, place the flask in cold water. When all the sulphuric acid has been added, place the flask in a dish containing water that has been heated to boiling. Allow to stand for five minutes, and then pour the contents of the flask into cold water. Nitrobenzene sinks to the bottom. Pour off the acid solution above it and add more water. Stir well, allow to settle, and pour off the water. Note the color and odor of nitrobenzene. Reaction: $C_6H_6 + HNO_3 = C_6H_5NO_2 + H_2O$. The sulphuric acid helps to remove the water.

(c) Aniline ($C_6H_5NH_2$).

Aniline is prepared by reducing nitrobenzene with iron and hydrochloric acid. Hydrogen is generated, which reacts as follows:

$$C_6H_5NO_2 + 6H = C_6H_5NH_2 + 2H_2O.$$

As it is troublesome to separate aniline from the products formed, it is better to use prepared aniline for the following experiments: Place a few drops of water in a test-tube and add an equal volume of conc. sulphuric acid. To this acid solution add a few drops of aniline, mix, and cool. Aniline sulphate is formed. Describe it.

Put 2 grms. of fresh bleaching-powder in 10 cc. of water. Stir well, allow to stand for five minutes, and filter. Dissolve a drop or two of aniline in 10 cc. of water and add this to the bleaching-powder solution. Note color. This is a test for aniline.

(d) Rosaniline.

Put a few drops of aniline with a few crystals of toluidine in a test-tube. Add a few crystals of mercuric chloride (corrosive sublimate) and heat for a moment over the open flame. Cool, add a drop of conc. hydrochloric acid, then 1 cc. of alcohol. Now dilute with water, and note the color imparted to the water.

References : Clk., pp. 329–337 ; R., pp. 408–411 ; S. and L., pp. 240–242.

APPENDIX.

INFORMATION FOR SCHOOLS NEEDING EQUIPMENT FOR TEACHING CHEMISTRY.

APPENDIX.

The following is a list of the apparatus called for by the manual. As it stands the list is complete for one student. It is desirable that all items marked * should be multiplied by the number of students who are expected to work at the same time.

The Denver Fire-clay Company, 3101–3141 Champa Street, Denver, Colo., will supply apparatus and chemicals at the prices given.

* Asbestos plate, $4'' \times 4''$	$0.05
Balloon, toy	.05
* Beakers, lipped, Nos. 0, 2, and 4	.31
* Blowpipe, plain, brass, $10''$	.06
* Bottles, 8 and 4 oz., wide mouth, 2 each	.15
* Burner, Bunsen, w. 2 feet hose	.32
* Burner, fishtail only	.10
* Burette clamp	.10
Corks, asst'd, 2 doz	.10
* Cork-borers, set of 3	.42
* Copper wire, 10 gauge, 3 feet	.05
* Crucible, porc., w. cover, $1\frac{1}{4}''$, R. M., No. 8	.09
* Evaporating-dish, porc., $3''$	.09
Files, round and three-cornered	.13
* Flasks, flat bottom, 250 and 500 cc	.18
* Funnel, $3''$	.07
* Gauze, wire, $4'' \times 4''$	.05
* Glass rod, $\frac{3}{16}$ diam., $9''$ long	.05

* Glass plate, window-glass, 3" × 3"............$0.05
Graduate, 50 cc. cylinder........................ .23
* Leaden dish, 2" diam........................... .08
Magnet, horseshoe, 2"............................ .05
Magnifying-glass, 1 glass, rubber case.......... .37
Mortar and pestle, W. W., 4 oz.................. .19
* Pincers, 4"..................................... .05
* Picture-wire, 6"......................No charge.
Platinum wire, 24 gauge, 3"..................... .24
Platinum-foil, med., 1" × 1".................... .60
* Pneumatic trough, 5" × 7" × 10"............... .84
Reagent bottles, set of eight, 4 oz.:

$\left.\begin{array}{l}\text{Hydrochloric acid conc.}\\ \text{Hydrochloric acid dil.}\\ \text{Sulphuric acid conc.}\\ \text{Sulphuric acid dil.}\\ \text{Nitric acid conc.}\\ \text{Nitric acid dil.}\\ \text{Sodium hydrate.}\\ \text{Ammonia.}\end{array}\right\}$ .88

* Retort, 50 cc. with glass stopper.............. .20
* Ring-stand, w. 2 rings......................... .34
* Test-tube rack for 6 tubes, w. pins............ .25
* Test-tubes, 16 × 160 mm., 1 doz................ .16
* Test-tube, one, 25 × 200 mm .... .03
* Thistle-tube, 14"............................... .06
* Tube, rubber, ¼", 6 inches .... .05
* Tubes, soft glass, ⅛" i. d., 5 tubes............ .10
* Tube, hard glass, ¼" o. d. .... .05
* Triangle for 1¼" crucible, pipe-stem........... .04
* Watch-glasses, 1½"............................. .05
Alcohol-lamp, 2 oz............................... .12

In addition to the apparatus just listed a number of bottles of various sizes, and a few large beakers and evaporating-dishes, should be included in the order. These articles will be needed in making solutions of reagents.

The laboratory should also be provided with a balance accurate to .05 grm., a set of weights, an apparatus for distilling water, and a water-bath.

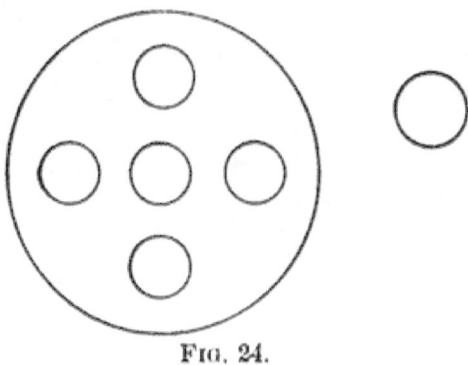

FIG. 24.

In case the building is heated by steam, water condensed in the steam-pipes may usually be obtained from the engineer. For laboratories not supplied with steam or with running water, a "sanitary" still is to be recommended. Information may be obtained from A. R. Bailey Mfg. Co., 54 Maiden Lane, N. Y.

A water-bath may be constructed out of a pan covered with a sheet-iron plate as shown in Fig. 24. The small circles represent holes about two inches in diameter for three-inch evaporators. The circle to the right represents a disk about three inches in diameter. Five of these should be provided for covers.

The following list of chemicals is estimated on the basis of ten students. As, however, it is not profitable

APPENDIX.

* Glass plate, window-glass, 3" × 3"	$0.05
Graduate, 50 cc. cylinder	.23
* Leaden dish, 2" diam.	.08
Magnet, horseshoe, 2"	.05
Magnifying-glass, 1 glass, rubber case	.37
Mortar and pestle, W. W., 4 oz	.19
* Pincers, 4"	.05
* Picture-wire, 6"	No charge.
Platinum wire, 24 gauge, 3"	.24
Platinum-foil, med., 1" × 1"	.60
* Pneumatic trough, 5" × 7" × 10"	.84
Reagent bottles, set of eight, 4 oz.:	

> Hydrochloric acid conc.
> Hydrochloric acid dil.
> Sulphuric acid conc.
> Sulphuric acid dil.   .... .88
> Nitric acid conc.
> Nitric acid dil.
> Sodium hydrate.
> Ammonia.

* Retort, 50 cc. with glass stopper	.20
* Ring-stand, w. 2 rings	.34
* Test-tube rack for 6 tubes, w. pins	.25
* Test-tubes, 16 × 160 mm., 1 doz	.16
* Test-tube, one, 25 × 200 mm	.03
* Thistle-tube, 14"	.06
* Tube, rubber, $\frac{1}{4}$", 6 inches	.05
* Tubes, soft glass, $\frac{1}{8}$" i. d., 5 tubes	.10
* Tube, hard glass, $\frac{1}{4}$" o. d.	.05
* Triangle for $1\frac{1}{4}$" crucible, pipe-stem	.04
* Watch-glasses, $1\frac{1}{2}$"	.05
Alcohol-lamp, 2 oz	.12

In addition to the apparatus just listed a number of bottles of various sizes, and a few large beakers and evaporating-dishes, should be included in the order. These articles will be needed in making solutions of reagents.

The laboratory should also be provided with a balance accurate to .05 grm., a set of weights, an apparatus for distilling water, and a water-bath.

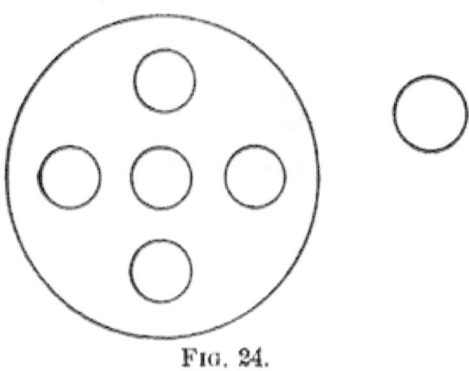

Fig. 24.

In case the building is heated by steam, water condensed in the steam-pipes may usually be obtained from the engineer. For laboratories not supplied with steam or with running water, a "sanitary" still is to be recommended. Information may be obtained from A. R. Bailey Mfg. Co., 54 Maiden Lane, N. Y.

A water-bath may be constructed out of a pan covered with a sheet-iron plate as shown in Fig. 24. The small circles represent holes about two inches in diameter for three-inch evaporators. The circle to the right represents a disk about three inches in diameter. Five of these should be provided for covers.

The following list of chemicals is estimated on the basis of ten students. As, however, it is not profitable to purchase in too small quantities, many of the substances are listed in amounts sufficient for several classes of this number. Items marked * are not included in the estimate furnished by the supply house.

CHEMICALS.

Acetic Acid, No. 8, and bot., $\frac{1}{4}$ lb..............	$0.07
*Alcohol. Procure revenue-free if used in quantity.	
Alum, potassium, $\frac{1}{2}$ lb........................	.12
Aluminium, foil, No. 34, 1 oz	.06
Aluminium Sulphate, $\frac{1}{4}$ lb.....................	.05
Ammonia, aqua, and bot., 5 lbs.................	.75
Ammonium Carbonate and bot., $\frac{1}{4}$ lb............	.11
Ammonium Chloride, gran., 2 lbs...............	.22
Ammonium Molybdate and bot., 1 oz...........	.20
Ammonium Nitrate, $\frac{1}{2}$ lb......................	.10
*Ammonium Sulphide. To be made in the laboratory.	
Animal Charcoal, gran., $\frac{1}{2}$ lb...................	.10
Aniline Oil and bot., 1 oz......................	.10
Antimony, powdered, 2 oz......................	.05
Arsenic, metallic, 1 oz.........................	.05
Arsenic Trioxide, com., and bot., 1 oz...........	.05
Barium Hydrate and bot., 2 oz	.18
Barium Chloride, in cart., $\frac{1}{4}$ lb.................	.10
Benzene, C_6H_6, $\frac{1}{4}$ lb.........................	.15
Bleaching-powder, small can....................	.09
Blue-print Paper, 4×5, 25 sheets, 1 pkge........	.15
Bone-ash, $\frac{1}{4}$ lb...............................	.05
Borax, 1 lb...................................	.10
Bromine and 2 bots. and 2 cans, 2 oz............	.40
Beeswax, 2 oz	.06
Cadmium, sticks and bot., $\frac{1}{2}$ oz...............	.15

APPENDIX.

Cadmium Chloride and bot., 1 oz	$0.30
Calcium Carbide, 1 lb	.28
Calcium Chloride, gr., and bot., 1 lb	.26
Calcium Fluoride, powdered, 1 lb	.05
Calcium Hypophosphite, ¼ lb. in bot	.42
* Calico, red.	
* Candles.	
Carbon Disulphide and can, 1 lb	.19
Chalk, precipitated, ½ lb	.07
Charcoal, 10 sticks	.25
Chloroform and bot., ¼ lb	.25
Chrome Alum, ¼ lb	.05
Chromic Acid and bot., ¼ lb	.19
Cobaltous Chloride and bot., 1 oz	20
Cobaltous Nitrate and bot., 1 oz	52
Copper turnings, 1 lb	.35
Copper Oxide, powder, ½ lb	.25
Copper Sulphate, ½ lb	.05
* Cotton, absorbent.	
Ferric Chloride and bot., 2 oz	.09
Ferrous Sulphate and bot., ¼ lb	.08
Filter-paper, gray, 25 sheets, 12 × 17	.18
Filter-paper, white, 11 cm. diam	.13
* Gasoline.	
Grape-sugar, commercial, 2 lbs	.19
Hydrochloric Acid and bot., 6 lbs	.85
Iceland Spar, 1 oz	.10
Indigo, 1 oz	.07
Iodine and bot., ½ oz	.20
Iron, reduced by hydrogen, ¼ lb	.45
Iron, filings, 1 lb	.05
* Kerosene.	
Lead foil, ¼ lb	.10
Lead Acetate and bot., 2 oz	.20

Lead Nitrate and bot., 2 oz $0.20
Lead Oxide, litharge, and bot., ¼ lb .... .15
Lead Oxide, minium, and bot., ½ lb .... .18
* Lime.
Lithium Chloride and bot., ½ oz .... .15
Litmus Paper, red and blue, large sheet, each... .06
Magnesium ribbon, 5 ft .... .10
Magnesium Sulphate, ¼ lb .... .05
Magnesium Oxide and bot., ¼ lb .... .40
Manganese Dioxide, 1 lb .... .06
Manganous Sulphate and bot., ¼ lb .... .20
* Marble or limestone.
Mercury and bot., ¼ lb .... .24
Mercuric Chloride and bot., 2 oz .... .20
Mercuric Oxide and bot., 2 oz .... .30
Methyl Alcohol and bot., ¼ lb .... .10
Nickel Sulphate, ¼ lb .... .17
Nitric Acid and bot., 7 lbs 1.02
Oxalic Acid, 1 lb .... .13
Paraffine, 2 oz .... .05
Phosphorus, yellow, and bot. and can, 2 oz ..... .38
Phosphorus, red, and bot., 2 oz .... .45
Phosphoric Acid, ortho, and bot., 2 oz .... .10
Potassium, metal, and bot. and can, 1 oz 1.70
Potassium Carbonate, dry, and bot., 2 oz .... .06
Potassium Bichromate, ½ lb .... .09
Potassium Bromide, ¼ lb .... .18
Potassium Chromate, ¼ lb .... .15
Potassium Hydrate, sticks, pure by alcohol, 1 lb.,
   and bot. .... .53
Potassium Ferrocyanide and bot., 1 oz .... .08
Potassium Ferricyanide and bot., 1 oz .... .10
Potassium Iodide and bot., ¼ lb .... .74
Potassium Permanganate, ½ lb .... .19

Rochelle Salt, ¼ lb. $0.09
* Salt.
* Soap.
Silver Nitrate and bot., 1 oz.. .... .50
Sodium, metallic, and container, 1 oz............ .35
Sodium Bicarbonate, ¼ lb. .... .05
Sodium Carbonate, dry powdered, ¼ lb. .... .05
Sodium Carbonate, crystals, ¼ lb................ .05
Sodium Hydrate, sticks, pure by alcohol, 1 lb., and
    bot..... .... .50
Sodium Nitrate, coml., 1 lb..................... .11
Sodium Phosphate (di), ¼ lb..................... .10
Sodium Silicate, water-glass, ½ lb., and bot...... .10
Sodium Sulphate, ½ lb........................... .05
Sodium Sulphide and bot., 1 oz. .... .08
* Starch.
Strontium Chloride, ¼ lb........................ .10
* Sugar.
Sulphur, flowers, 1 lb..... .... .05
Sulphur, roll, 1 lb......... .... .05
Sulphuric Acid, c. p., and bot., 9 lbs............ 1.15
Tartaric Acid, ½ lb. .... .23
Tartar Emetic and bot., 1 oz.................... .08
Tin, granulated, 2 oz............................ .06
* Turpentine.
Toluidine (para), crystals, and bot., 1 oz......... .32
* Yeast.
Zinc, granulated, 2 lbs........................... .70
Zinc Oxide, by wet process, 2 oz................ .10

    Except when specified chemicals are put up in pasteboard boxes. Bottles are furnished for all chemicals at $2.00 extra.

    The plan shown in Figs. 25 and 26 represents a

APPENDIX.

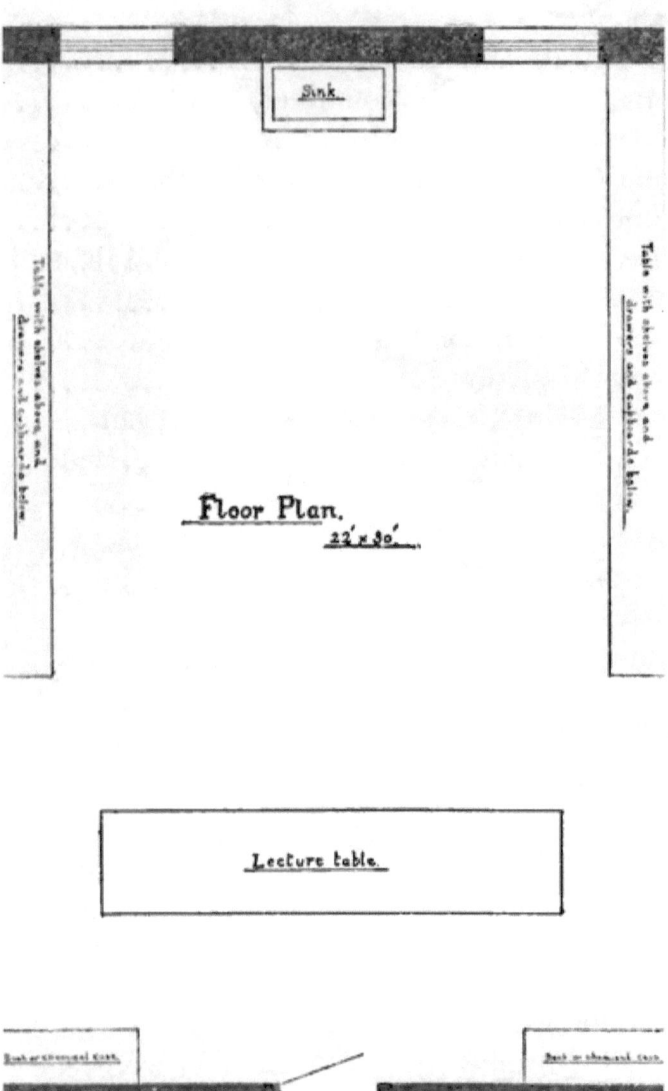

Fig. 25.

APPENDIX.

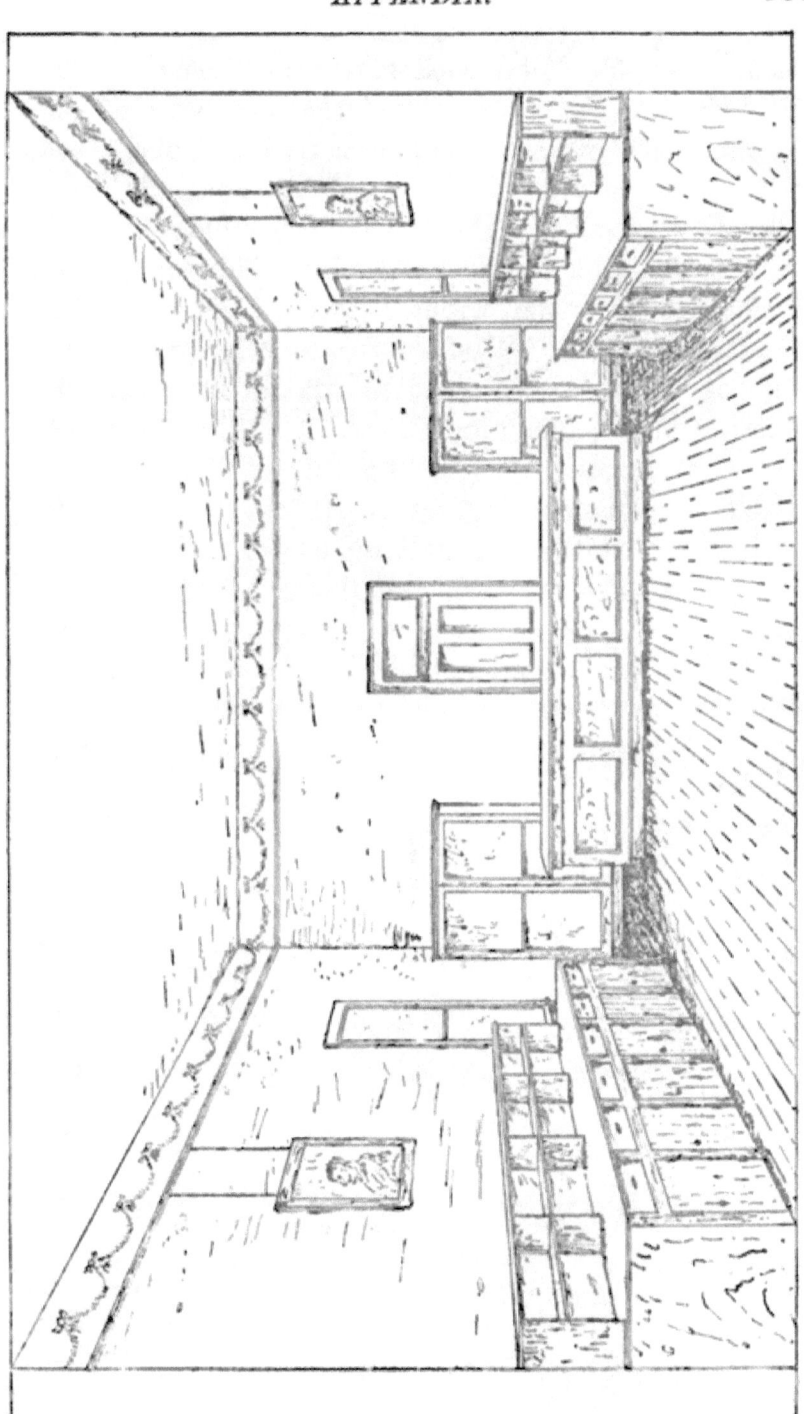

Fig. 26.

room 22' × 30' fitted with two work-tables 18' long, 38" high, and 30" wide. The lecture-table is 14' long, 33" high, 36" wide. The two cases for books, chemicals, and apparatus are 5' long, $6\frac{1}{2}$' high, and 18" deep, with about four shelves to each case, glass doors in front. The sink should be of enamelled iron or slate, preferably the latter, though costing somewhat more, size about 18" × 40", fastened to iron brackets on wall or cased up below with wood. The following material may be used for construction of cases, tables, etc.: For framing 2" × 4" scantling; doors in work-tables and panels in lecture-tables $\frac{5}{8}$" yellow-pine ceiling; drawer-fronts, shelving, cases, casing, etc., $\frac{7}{8}$" soft pine; tops of work and lecture tables $1\frac{1}{4}$" soft pine.

A laboratory fitted as above described will cost from $125.00 to $130.00, including varnish, etc.

The central part of the room has ample capacity for seating.

www.ingramcontent.com/pod-product-compliance
Lightning Source LLC
Chambersburg PA
CBHW022128160426
43197CB00009B/1196